Tony Anastasio
10 Brownle
Dukinfie
Cheshi
SK16 4TZ.

Motor Vehicle Technology

Associated Studies 1

by the same author

Fundamentals of Motor Vehicle Technology
(with F.W. Pittuck)

Motor Vehicle Technology: Associated Studies 2

Fundamentals of Automotive Electronics

MOTOR VEHICLE TECHNOLOGY

Associated Studies 1

V. A. W. Hillier T.Eng. (CEI), FIMI, AMIRTE

Senior Lecturer in Automobile Engineering,
Croydon College

Stanley Thornes (Publishers) Ltd

First published in 1974 by Hutchinson Education
Reprinted 1979, 1981, 1982, 1984, 1988

Reprinted 1990 by
Stanley Thornes (Publishers) Ltd
Old Station Drive
Leckhampton
CHELTENHAM GL53 0DN

Reprinted 1992

ISBN 0 7487 0383 7

Printed and bound in Great Britain at
Scotprint Ltd, Musselburgh

CONTENTS

PREFACE

Since 1965 Great Britain and many other countries have been changing to a metric system of units. At this period of time many industries have completed their metrication programme whereas others have just begun; the motor industry is in the latter group. Many difficulties face the manufacturers, but in the next few years most companies will adopt the International System of Units or as it is commonly called the SI system.

Unless careful preparations are made, the service and repair section of the motor industry will experience many problems owing to the wide range and age of the vehicles serviced. During the change-over period both the old Imperial system and the new SI system will be used side-by-side and this will make extra demands on all concerned. However, it is generally realised that new techniques must be mastered and old standards must be modernised if the country is to prosper.

The book has been written to provide the reader with the basic knowledge of subjects related to motor vehicle technology. Since SI units are used throughout the book it is hoped that the advantages associated with a coherent, universal system of units will be apparent.

Although the basic units used in the book are SI, at each stage the Imperial unit is also stated, together with the method for converting from the old system to the new system. This provision will enable the reader to apply data from any vehicle service manuals which uses non-SI units.

Material for the book has been based on the syllabus entitled 'Motor Vehicle Craft Studies Part 1' which is published by the City and Guilds of London Institute and Regional Examining Bodies. The author's wide experience of this examination has enabled him to indicate the depth of treatment required for each topic.

Questions in the national examinations are set in the multiple-choice, objective pattern, so in this book the use of over 200 multi-choice questions together with advice on the method of answering this type of examination paper should help the reader to acquire the necessary technique and experience.

Metric systems place greater emphasise on decimals so this subject has been stressed. Conversely, new systems have made the vulgar fraction redundant, so treatment of this topic has been restricted to necessities such as $3\frac{1}{7}$.

Much of the book has been applied to motor vehicles and it acts as a companion to *Fundamentals of Motor Vehicle Technology*.

When a cross reference is shown the title is abbreviated to 'F of M.V.T.'

In general the recommendations made by the British Standards Institute have been followed.

V. A. W. Hillier

1 CALCULATIONS

In the modern world it is necessary to use simple mathematics in order to handle everyday problems. Although we may employ complicated machines to assist us in our work it is still essential that we can manipulate simple numbers.

Often the success in our chosen career is governed by our ability to handle a set of figures, since other studies often depend on this basic subject.

The subjects covered in this section have been covered at school, so all that is necessary is to revise and consolidate each section. Many items recur at a later stage so it is essential that each stage is fully understood before proceeding.

1.1 Decimal system—Introduction

Decimal indicates a system which is based on the number 10. Applying this to whole numbers – the number 574 means the sum of:

5 hundred	or	500
7 tens	or	70
4 whole units	or	4
		574

Each unit could be divided into smaller parts to form a fraction of a whole unit. To separate the fraction from the whole unit, a decimal point (a dot) is used, e.g. 574·28.

Since the system is based on ten, then the first place after the decimal point represents tenths of a whole unit and the second place the hundredths of a whole unit. So the fractional part of the number 574·28 consists of

2 tenths of a whole unit	or	$\frac{2}{10}$
8 hundredths of a whole unit	or	$\frac{8}{100}$

This can also be shown by a diagram

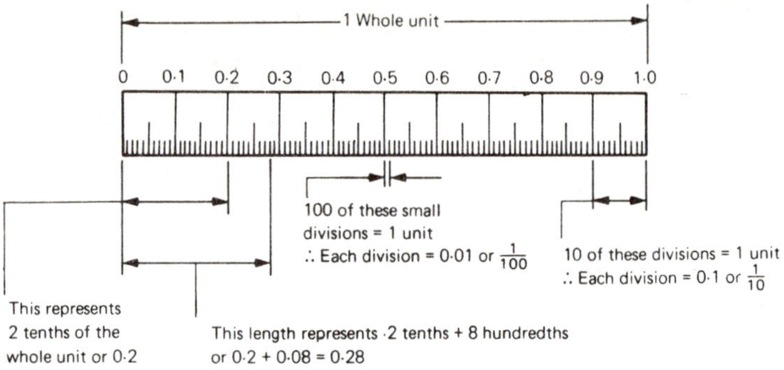

Fig. 1.1

Using this system it is possible to show very small fractions in a similar way to that used to indicate large values or whole numbers. To the left of the decimal point each column increases to give tens, hundreds, thousands, etc. whereas moving to the right away from the decimal point; each column is a tenth of the previous column and gives tenths, hundredths, thousandths, etc. When the number is less than unity (1) a zero should precede the point to draw attention and avoid errors. So the fraction is written as:

0·28 and not ·28

In the past large numbers have been grouped in thousands and a comma has been used, e.g. 5,623,428·62. The practice on the Continent is to use a comma for a decimal point so to avoid a misunderstanding it is recommended that the digits are spaced out in groups of three and commas not used.

Examples

1. 5 623 428·62
2. 7 952·864 205
3. 0·000 052

Multiplication and division by multiples and submultiplies of 10

Often it is necessary to either multiply or divide by some multiple of 10. This is performed by moving the decimal point:

When multiplying – **Move point to Right** – since answer must be bigger
When dividing – **Move point to Left** – since answer must be smaller.

The amount that the point is moved depends on the size of the multiplying number in relation to 10.
To multiply by:

10 move point 1 place to the right
100 move point 2 places to the right
1 000 move point 3 places to the right
1 000 000 move point 6 places to the right

Division is similar except point is moved to the left.

Examples

$3\,250{\cdot}462 \times 1000 = 3\,250{\cdot}\overset{1\,2\,3}{462} = 3\,250\,462{\cdot}$

$0{\cdot}625 \times 100 = 0{\cdot}\overset{1\,2}{62}5 = 62{\cdot}5$

$0{\cdot}000\,5 \times 1\,000\,000 = 0{\cdot}\overset{1\,2\,3}{000}\,\overset{4\,5\,6}{500} = 500{\cdot}$

$323\,471{\cdot}2 \div 1000 = 323\,\overset{3\,2\,1}{471}{\cdot}2 = 323{\cdot}471\,2$

$620{\cdot}4 \div 100 = 6\overset{2\,1}{20}{\cdot}4 = 6{\cdot}204$

$62\,052{\cdot}4 \div 1\,000\,000 = \overset{6\,5\,4}{062}\,\overset{3\,3\,1}{052}{\cdot}4 = 0{\cdot}062\,052\,4$

This layout may be simplified by using 'powers of 10' instead of writing the full value. The value 100 is also given by 10 x 10 and this is written as 10^2 and spoken as ten squared or 'ten to the power of 2'. In a similar way:

$1000 = 10 \times 10 \times 10 = 10^3$ (ten cubed)
$1\,000\,000 = 10 \times 10 \times 10 \times 10 \times 10 \times 10 = 10^6$ (ten to the power of 6)

The number placed above the 10, (in these cases 3 and 6) is called an INDEX. Using this method it is also possible to express a fraction of 10 by using a minus sign so:

$0{\cdot}1 = 10^{-1}$
$0{\cdot}001 = 10^{-3}$
$0{\cdot}000\,001 = 10^{-6}$

Examples

$27{\cdot}242 \times 10^3 = 27\,242$ (point moved 3 places to right)
$0{\cdot}000\,52 \times 10^6 = 520$ (point moved 6 places to right)

When a value is to be multiplied by a fraction of 10 then the answer will be smaller:

(a) $0{\cdot}1 \times 27 = 2{\cdot}7$

i.e. $\frac{1}{10}$th of 27 = 2·7

(b) $0{\cdot}001 \times 460 = 0{\cdot}460 = 0{\cdot}46$

i.e. $\frac{1}{1000}$th of 460 = 0·46

This is easier if it is expressed as a power of 10:

(a) $27 \times 10^{-1} = 2{\cdot}7$ (point moved 1 place to left)
(b) $460 \times 10^{-3} = 0{\cdot}46$ (point moved 3 places to left)

Rule When multiplying by a value which is:

(a) greater than unity (positive index) move point to the right
(b) smaller than unity (negative index) move point to the left.

This 'power of ten' method of presentation may be used to reduce a number to a shorter form.

Examples

17 600 000	may be written as	$17{\cdot}6 \times 10^6$ (or $1{\cdot}76 \times 10^7$)
32 650	" " " "	$32{\cdot}65 \times 10^3$
0·000 052	" " " "	52×10^{-6}
0·006 21	" " " "	$6{\cdot}21 \times 10^{-3}$

It is preferable to keep the index a multiple of 3 to suit the SI units given on page 14.

Instances occur where a large number has to be multiplied by a fraction.

Example

$$4000 \times 0{\cdot}07$$

This could be written as

$$40 \times 100 \times 0{\cdot}07$$

Multiplying 0·07 by 100 gives 7·0

so $\quad 4000 \times 0{\cdot}07 = 40 \times 7 = 280$

Example

28 300 x 0·005	(0·005 has been multiplied by 1000 to remove
28·3 x 5	the fraction. To keep the product the same, the
141·5	number 28 300 must be divided by 1000)

Example

40 000 x 0·0127

4 3 2 1   1 2 3 4
40 000 x 0·0127   (decimal point moved 4 places)

4·0 x 127·0

508

Exercises 1.1

Find the value of:

1.	16·2	x	1000
2.	0·0461	x	1000
3.	4·216 31	x	1 000 000
4.	0·102	x	100
5.	0·76	x	1000
6.	52·5	÷	100
7.	32 761	÷	1000
8.	52 000 000	÷	1 000 000
9.	43·72	÷	1000

10.	0·504	÷	100
11.	43·1	x	10^3
12.	0·0521	x	10^3
13.	0·000 62	x	10^6
14.	0·23	x	10^2
15.	0·6	x	10^1
16.	13·4	x	10^{-3}
17.	232·6	x	10^{-3}
18.	17 415	x	10^{-6}
19.	70 000 000	x	10^{-6}
20.	5·2	x	10^{-2}

1.2 Decimal system— Addition, subtraction, multiplication and division

Addition of decimals

Example 1

Add 24·635, 4·62 and 0·375.

```
24·635
 4·62
 0·375
------
29·630
```

Arrange the numbers under each other making sure the decimal points are *all in line.*

Add each column in the normal way.

Example 2

Add 4·003, 16·742 and 8·352.

```
 4·003
16·742
 8·352
------
29·097
```

Subtraction of decimals

Example 1

Subtract 24·672 from 49·894.

```
49·894
24·672
------
25·222
```

As when adding, the decimal points must be placed in line.

Example 2

From 23·474 take 6·826.

```
23·474
 6·826
------
16·648
```

Answer to a subtraction problem may be checked by adding second and third lines: 6·826 + 16·648 = 23·474.

Multiplication of decimals

When multiplying decimals together, first multiply in the normal way and then determine the position of the decimal point.

Example 1

2·54 x 6·1

```
   2·54
    6·1
  -----
  15240
    254
  -----
  15494
  -----
```

There are three decimal places in this problem; 0·54 and 0·1.
To position the point count from the right the number of decimal places. In this case there are three places, so the point is placed between the 5 and 4.

Answer 15·494.

Example 2

14·61 x 0·005

```
  14·61
   ·005
  -----
   7305
  -----
```

There are five decimal places and only four digits in the result, so 0 must be placed in front of the 7 to make up five places.

Answer 0·073 05.

Example 3

27·361 x 3·142

```
    27·361
     3·142
  --------
  82083000
   2736100
   1094440
     54722
  --------
  85968262
  --------
```

Answer 85·968262 or 85·968 correct to three decimal places.

Before dividing decimals the denominator is changed to a whole number.

Division of decimals

Example 1

16·36 ÷ 0·04

This can be written as

$$\frac{16{\cdot}36}{0{\cdot}04}$$

N.B. Decimal points placed under each other.

Multiplying numerator and denominator by 100 to change the denominator to a whole number

$$\frac{1636\cdot}{4\cdot} = 409$$

N.B. To multiply by 100 shift the decimal point two places to the right.

Example 2

107·44 ÷ 3·4

$$\frac{107\cdot 44}{3\cdot 4} = \frac{1074\cdot 4}{34\cdot}$$

In this case the decimal point need only be moved one place; i.e. multiplied by 10.

```
      31·6
34)1074·4
   102
    ---
     54
     34
     --
     204
     204
```

The decimal point in the answer must be placed above the point in the number 1074·4.

Answer 31·6.

Example 3

14·1 ÷ 3·142 correct to three decimal places.

$$\frac{14\cdot 1}{3\cdot 142} = \frac{14100\cdot}{3142\cdot}$$

```
         4·4875
3142)14100
     12568
     -----
      15320
      12568
      -----
       27520
       25136
       -----
        23840
        21994
        -----
         18460
         15710
```

0's can be brought down until the answer works out or the required number of places are obtained.

If the fourth place is 5 or over the third place is raised by 1.
The answer to this example is 4·488.

Example 4

$0{\cdot}084 \div 23{\cdot}2$ correct to three decimal places.

$$\frac{0{\cdot}084}{23{\cdot}2} = \frac{0{\cdot}84}{232{\cdot}}$$

```
       0·0036
232)0·840
      696
      1440
      1392
```

Answer 0·004.

Converting a vulgar fraction to a decimal fraction

A vulgar fraction is an alternative method to the decimal system for expressing a part or parts of a whole unit. If a whole unit is split into say four parts then each separate part will form a quarter and is written as ¼. When three of these parts are considered then the amount will be ¾. A vulgar fraction such as ¾ consists of:

$$\frac{\text{Numerator}}{\text{Denominator}}$$

The number on the lower line indicates the parts into which the whole unit is broken; the numerator states the number of these parts present.

When a vulgar fraction has a denominator which is a multiple of 10 then it can be converted to a decimal by placing the numerator in the appropriate column.

Example 1

$$\frac{3}{10} = 0{\cdot}3$$

$$\frac{9}{100} = 0{\cdot}09$$

$$\frac{1}{1000} = 0{\cdot}001$$

$$4\frac{17}{1000} = 4{\cdot}017$$

$$3\frac{13}{100} = 3{\cdot}13$$

$$2\frac{137}{1000} = 2{\cdot}137$$

Denominators other than those based on 10 are converted by dividing the denominator into the numerator.

Example 2

Convert $\frac{3}{8}$ to a decimal.

$$8\overline{)3000}$$
$$\cdot375$$

0's are added after the 3 until a remainder is eliminated.

So $\frac{3}{8} = 0{\cdot}375$

Example 3

Convert $4\frac{7}{8}$ to a decimal.

Now $\frac{7}{8}$ is converted as before:

$$8\overline{)7000}$$
$$\cdot875$$

So $4\frac{7}{8} = 4{\cdot}875$.

Example 4

Convert $\frac{1}{7}$ to a decimal.

$$7\overline{)100000}$$
$$\cdot14285$$

It will be seen that this number does not work out – a remainder exists after working it to a large number of places. The greater the number of places following the decimal point, the more accurate will be the result. For general work, three decimal places are normally sufficient.

If three decimal places are required, the conversion is calculated to four places and if the number in the fourth decimal place is 5 or over, the number in the third decimal place is raised by 1.
So $\frac{1}{7}$ correct to 3 decimal places = 0·143.

Example 5

Convert $\frac{1}{3}$ to a decimal.

$$3\overline{)1000}$$
$$\cdot333$$

Again this does not work out. Instead of repeating a number many times, a dot is placed over the *recurring number* or numbers, so $\frac{1}{3} = 0{\cdot}\dot{3}$.

The following fractions are often used, so they should be memorised:

$\frac{1}{32} = 0{\cdot}031\ 25$	$\frac{1}{16} = 0{\cdot}0625$	$\frac{1}{8} = 0{\cdot}125$
$\frac{1}{4} = 0{\cdot}25$	$\frac{3}{8} = 0{\cdot}375$	$\frac{1}{2} = 0{\cdot}5$
$\frac{5}{8} = 0{\cdot}625$	$\frac{3}{4} = 0{\cdot}75$	$\frac{7}{8} = 0{\cdot}875$

Cancellation

This method enables smaller numbers to be obtained, but special care must be exercised when it is used.

Example 6

$$\frac{\overset{11}{\cancel{22}} \times \overset{1}{\cancel{\overset{5}{\cancel{35}}}} \times \overset{1}{\cancel{3{\cdot}2}}}{\underset{1}{\cancel{7}} \times \underset{\underset{3}{\cancel{6}}}{\cancel{\underset{30}{\cancel{960}}}}} = \frac{11}{3} \simeq 3{\cdot}7$$

N.B. The 7 and 35 are divided by 7,
3·2 and 960 are divided by 3·2
5 and 30 are divided by 5
22 and 6 are divided by 2

This example shows that a fraction keeps its value provided the same number is divided into the top line as the bottom line.

Normally it is unnecessary to write 1 after 'cancelling-out' a number since if the number 'goes-out' then the value 1 must apply.

Example 7

$$\frac{33 \times 0{\cdot}024}{3\frac{1}{7} \times 14 \times 5} = \frac{33 \times 0{\cdot}024}{\frac{22}{7} \times 14 \times 5}$$

$$= \frac{\cancel{7} \times \overset{3}{\cancel{33}} \times \overset{0{\cdot}006}{\cancel{0{\cdot}024}}}{\underset{2}{\cancel{22}} \times \underset{2}{\cancel{14}} \times 5}$$

To divide by a vulgar fraction the fraction is inverted.

$$= \frac{0{\cdot}018}{5} = 0{\cdot}0036$$

Exercises 1.2

Convert to a decimal fraction:

1. $\frac{7}{10}$ 2. $\frac{3}{32}$ 3. $\frac{1}{6}$ 4. $\frac{1}{15}$

Add

5. 6·4; 3·12; 4·05.
6. 17·2; 4·716; 0·032; 61·1.
7. 605; 43·614; 0·0062; 0·4; 3·142.
8. 23·04; 0·051; 4·301; 5·32; 0·63; 4·193.

Subtract

9. 23·4 from 66·9.
10. 3·27 from 17·56.
11. 2·04 from 6·7.
12. 3·375 from 17·031.

Multiply

13. 54·3 by 0·7.
14. 3·27 by 2·45.
15. 9·72 by 0·0358.
16. 17·62 by 379·1.

Divide

17. 103·2 by 0·06.
18. 972·4 by 2·6.
19. 23·6 by 3·74 correct to three decimal places.
20. 0·0715 by 4·215 correct to three decimal places.

1.3 SI units

If you have travelled out of this country you will appreciate the difficulties that arise when you try to communicate with people living in others lands. Many of these difficulties would be avoided if the traveller was able to speak the language of the country, but this requires considerable effort, intelligence and practice. Consider the advantages that would be gained if we all spoke one common language. In the past the 'language difficulty' has applied to the various units which are used to represent such things as distances or quantities. Not only was it difficult to understand the units used in a country, but it often caused confusion – two examples of this are the 'ton' and the 'gallon'. An 'American ton' and 'American gallon' are quite different to their namesakes in Great Britain, and if France is included, then yet another value is obtained.

With the increase in trade between nations an international body was set up to make recommendations for suitable standards. This body is called the International Standards Organisation (ISO) and in Britain the organisation associated with ISO is the British Standards Institute (BSI). At an international conference in 1960 it was recommended that everyone should use a metric system of measurement called 'Systeme International d'Unites' and this is normally abbreviated to SI in all languages. The system was based on six primary units:

Unit	*Symbol*	*Quantity*
metre	m	length
kilogram	kg	mass
second	s	time
ampere	A	electric current
kelvin	K	temperature
candela	cd	luminous intensity

Metre is the unit of length

Distances are expressed in this unit. Originally the metre represented a part of the distance between the earth's pole and equator, and a bar of platinum was kept in Paris to form the standard length. Nowadays the need for greater accuracy has demanded a more stable standard, so in 1960 the metre was redefined as the length obtained by using the wavelengths of a certain type of light.

Kilogram is the unit of mass

The standard kilogram is a certain cylindrical block of metal kept in France and was originally intended to represent a mass of 1000 cubic centimetres of water. The term mass is the quantity of matter in a body.

When a mass of 1 kg is placed on one side of a simple balance, it will serve as a standard to compare other substances. For example, if 1 kg of grease is required, then this could be 'weighed' on a balance until the two masses are equal. When this occurs the earth's pull of gravity will give a force on the standard mass which will be the same as that acting on the grease.

Second is the unit of time

You are aware that a day has 24 hours, one hour has 60 minutes, and one minute has 60 seconds. Before 1967 astronomical observations were used to check the special clocks housed at Greenwich, but since that date the second has been redefined as the time given by the action of a certain element in an atomic clock.

The other base units will be discussed at a later stage in the book.

Often the recommended unit is either too large or too small, e.g. the distance between London and Manchester is 297 000 metres and a typical gap for a sparking plug is 0·000 635 metres. To avoid this cumbersome presentation, multiples and submultiples of SI units are used and the following table shows the common prefixes used in our type of work.

Unit multiplier	*Prefix*	*Symbol*
One million or 10^6	mega	M
One thousand or 10^3	kilo	k
One hundredth or 10^{-2}	centi	c
One thousandth or 10^{-3}	milli	m
One millionth or 10^{-6}	micro	μ

Using this system the distance between London and Manchester is given as 297 kilometres and the sparking plug gap as 0·635 millimetre or in even shorter form as 297 km and 0·635 mm respectively. Since common practice is to use symbols, then both SI and BSI recommendations should be followed. The symbol 'm' already represents 'milli' and 'metre' so it is obvious that it cannot be used for a minute of time. In cases like this, the non SI abbreviations are changed and this is the reason why expressions such as R.P.M. should now be stated as rev/min.

The forementioned prefixes can be applied to all SI units.

Examples

220 613 grams	=	220·613 kg
0·02 second	=	20 ms
0·0075 kilometre	=	7·5 m
700 milliamperes	=	0·7 A

These examples also show why the letter 's' should not be added to the symbol to form a plural. Where confusion is likely the unit should be written in full — in this case the plural can be used where appropriate.

Having considered these basic features of SI we can look into some other aspects. The table of prefixes is based on the metric system, which, as applied to the length unit, is as follows:

10 millimetres (mm)	=	1 centimetre	(cm)
10 centimetres	=	1 decimetre	(dm)
10 decimetres	=	**1 metre**	(m)
10 metres	=	1 decametre	(dam)
10 decametres	=	1 hectometre	(hm)
10 hectometres	=	**1 kilometre**	(km)

Inspection of this table shows that some units have not been included in previous sections of this chapter. This is because in SI, certain units are preferred and these are shown in the table.

Derived units

Many other units can be derived from the six primary units to form a system which is all linked together. Thus if two units are multiplied together the resultant quantity forms a new unit. Some of these are considered at this stage; others are mentioned when they occur in the book.

Area

Unit of area is the square metre (m^2).

Since there are 100 centimetres in a metre, there will be 100 x 100 square centimetres in 1 square metre. So

$$10\,000 \text{ cm}^2 = 1 \text{ m}^2$$

Volume

Unit of volume is the cubic metre (m^3).

Using the length measurement as a basis it will be seen that

$$1 \text{ cm}^3 = 10 \times 10 \times 10 \text{ mm}^3$$

so

$$1 \text{ cm}^3 = 1000 \text{ mm}^3$$

In the past the symbol for a cubic centimetre was 'cc' and in many cases the total swept volume (swept volume of one cylinder X number of cylinders) of an engine was stated in cc (e.g. 250 cc).

Capacity

Unit of capacity is the litre (l).

Originally this was stated as the volume occupied by a mass of 1 kg of pure water measured under stated conditions, but due to slight errors the litre was redefined in 1964 as equal to 1000 cm^3. Although the litre is not a preferred SI unit, its general use makes it necessary to include it here.

Engine capacity is often stated in litres. Since 1000 cm^3 = 1 litre, an engine having a capacity of 1500 cm^3 would be called a $1\frac{1}{2}$ litre engine.

Mass

Unit of mass is the kilogram (kg).

This primary SI unit comes from the metric table:

1000 mg = 1 gram (g)
1000 g = 1 kg
1000 kg = 1 tonne (t) (To avoid misunderstanding it is advisable to write tonne in full)

Exercise 1.3

1. What is the SI unit of

(a) mass (b) time (c) length

2. Convert:

(a) 350 mm to m
(b) 6240 m to km
(c) 24 cm to mm
(d) 6 Mm to km
(e) 27 μm to mm
(f) 0·625 km to m
(g) 0·25 m to cm
(h) 0·03 mm to μm

(i)	0·7 m	to	mm
(j)	0·0102 cm	to	μm

3. Convert

(a)	3×10^3 m	to	km
(b)	$5{\cdot}2 \times 10^6$ m	to	Mm
(c)	$0{\cdot}003 \times 10^3$ m	to	mm
(d)	22×10^{-3} m	to	mm
(e)	3562×10^{-6} m	to	mm
(f)	220×10^3 g	to	kg
(g)	560×10^6 mg	to	kg
(h)	22×10^3 cg	to	g
(i)	$0{\cdot}51 \times 10^{-3}$ g	to	mg
(j)	$0{\cdot}02 \times 10^{-6}$ kg	to	mg

4. An engine uses oil at the rate of 3·17 litres per month. Estimate the total oil consumption for a period of 5·5 months.

5. During one month three quantities of petrol; 18·5 litre, 14·75 litre and 15·25 litre were supplied on credit to a customer. What will be the total charge for this petrol if the fuel costs 8 pence (£0·08) per litre?

6. A wheel rim has a nominal diameter of 356 mm and this rim is fitted with a tyre which holds the rim 103·37 mm from the road surface. What is the effective diameter of the wheel?

7. A piston of diameter 79·924 mm is fitted in a cylinder having a bore of 80 mm. What is the clearance between the piston and the cylinder?

8. The length of a carburetter control wire is 2·53 m. How many lengths of this control wire can be cut from a coil containing 100 m of new wire?

9. To repair a vehicle after a crash it is estimated that the time required by the Body, Electrical and Mechanical sections is 11·75, 1·6 and 3·5 hours respectively. What is the estimated cost for the repair if the labour charge of £3·50 per hour is made?

10. To repair an electrical component, three cables of lengths 0·84 m, 0·52 m and 1·25 m were used. Find the total length of cable used.

11. When a special friction lining wears down to 0·3 of its original thickness, it is recommended that the lining is replaced. How much wear is permissible if the new lining has a thickness of 16·41 mm?

12. The wheelbase of a car is 2·2 times the track of the vehicle. Calculate the track of the car in mm if the wheelbase is 2·64 m.

13. A vehicle has a mass of 1 tonne. Calculate the mass acting on the rear wheels if the load on the rear wheels is 0·4 of the total.

14. A mechanic is paid £0·64 per hour and receives time and a half for overtime. How much overtime payment does he receive if he works 3·6 hours over his normal time?

1.4 Averages and proportion

Averages

An average or mean is a number around which other numbers are grouped, i.e. an intermediate to several numbers.

Consider the capacity of three petrol tanks. If tank A holds 5 litres, B holds 7 litres and C holds 9 litres then the average capacity of these three tanks would be 7 litres. On taking into consideration a fourth tank of capacity 11 litres, the average would become 8 litres. The method used to find the average is to total up the various values and divide into this total the number of values.

In the case of the four tanks above:

$$\text{Average} = \frac{\text{total capacity of all tanks}}{\text{number of tanks}}$$

$$= \frac{5 + 7 + 9 + 11}{4} = \frac{32}{4} = 8 \text{ litres}$$

Example 1

Over a five-week period the wage of a mechanic was £14; £19; £17; £22 and £18. What was his average wage during this period?

$$\text{Average} = \frac{\text{total wages}}{\text{number of weeks}}$$

$$= \frac{14 + 19 + 17 + 22 + 18}{5}$$

$$= \frac{90}{5} = \pounds 18$$

This problem shows that if he received £18 every week then the total after five weeks should be the same as the total of the varying wage.

Example 2

During a period of six months the total distance covered by a car was 12 506 km. Determine the

(a) average distance covered per month
(b) average distance covered per week assuming the 6 month period represents 26 weeks.

(a) 12 506 is the total distance covered

$$\text{Average distance covered per month} = \frac{12\,506}{6} \triangleq 2084 \text{ km}$$

(The sign $\triangleq$ means 'is approximately equal to')

(b) $$\text{Average distance covered per week} = \frac{12\,506}{26} = 481 \text{ km.}$$

Example 3

The petrol consumption of a certain car, measured over various journeys is:

3·5 litres for a journey of 42 km
1·95 litres for a journey of 20 km
7·55 litres for a journey of 94 km

What is the average petrol consumption for the three journeys?

This type of problem is treated in a slightly different manner. In this case the total distance covered is divided by the total petrol consumption.

$$\text{Average consumption} = \frac{\text{total distance}}{\text{total petrol consumption}}$$

$$= \frac{42 + 20 + 94}{3{\cdot}5 + 1{\cdot}95 + 7{\cdot}55}$$

$$= \frac{156}{13}$$

$$= 12 \text{ km/litre}$$

Example 4

A car travels 8 km at 40 km/hour and then 6 km at 60 km/hour. What is the average speed for the whole journey?

$$\text{Time taken for 8 km} = \frac{8}{40} = 0{\cdot}2 \text{ hour}$$

$$\text{Time taken for 6 km} = \frac{6}{60} = 0{\cdot}1 \text{ hour}$$

$$\text{Total time taken} = 0{\cdot}2 + 0{\cdot}1 = 0{\cdot}3 \text{ hour}$$

$$\text{Total distance covered} = 8 + 6 = 14 \text{ km}$$

$$\text{Average speed} = \frac{\text{total distance}}{\text{total time}}$$

$$= \frac{14}{0{\cdot}3} = 46{\cdot}6 \text{ km/hour}$$

Ratio

Ratio means the relationship between two quantities of a similar form or family, i.e. the fraction or multiple the first is to the second.

The ratio of a length of 51 mm to a length of 2 m (2000 mm) is $\frac{51}{2000}$ or 51:2000, which is read as 51 to 2000. The form or family in this case is 'length'. Both quantities must be expressed in the same units in order to show the true relationship.

One common application of this subject is in respect to gearing. For example, suppose for every 10 turns of the engine, the propeller shaft rotates twice. The gearbox ratio would be $\frac{10}{2}$ or 10 : 2, which is reduced to its lowest form to give $\frac{5}{1}$ or 5 : 1. This ratio shows that 5 turns of the engine causes the propeller shaft to rotate once.

Example 1

Express the ratio of 2 litres to 50 millilitres

ratio $= \frac{2000}{50}$ since 1000 ml = 1 litre

$= \frac{\cancel{2000}^{40}}{\cancel{50}}$

$= \frac{40}{1}$ or 40 : 1

Example 2

Express the ratio of 45 km/hour to 50 m/s

45 km/h = 45 000 m/h

$= \frac{45\,000}{60 \times 60}$

$= \frac{\cancel{450}^{50}}{\cancel{36}_{4}} = 12{\cdot}5$ m/s

Ratio = 12·5 : 50 or doubling both values to clear the fraction

= 25 : 100

= 1 : 4 or $\frac{1}{4}$

In these examples it will be seen that when the ratio is stated as a vulgar fraction the first term forms the numerator and the second term forms the denominator.

Example 3

To obtain sufficient protection of the cooling system 1 part of a certain brand of anti-freeze has to be mixed with 4 parts of water, i.e. the ratio of anti-freeze to water is 1 : 4. How much anti-freeze is required for a cooling system having a capacity of 5 litres?

If ratio is 1 : 4 then 1 part is anti-freeze, 4 parts are water.

Total = 5 parts and represents capacity of system.

Thus $\frac{1}{5}$ of system's capacity is the quantity of anti-freeze and $\frac{4}{5}$ of system's capacity is the quantity of water.

So

Quantity of anti-freeze required = $\frac{1}{5}$ of 5 litres

= 1 litre

This example shows that the whole quantity may be found by adding the ratios. If three quantities were given; e.g. 3 : 4 : 5, then the fraction of the whole represented by each would be $\frac{3}{12}$; $\frac{4}{12}$; $\frac{5}{12}$.

Example 4

One type of solder consists of lead and tin in the ratio 6:4. How much lead is used in 30 kg of solder?

$$\frac{6}{10} \text{ of the solder is lead}$$

$$\text{So quantity of lead in 30 kg of solder} = \frac{6}{10} \times 30 = 18 \text{ kg}.$$

Proportion

The term 'proportion' is used to show the variation of one quantity in respect to another independent quantity.

Consider a vehicle moving at a constant speed. The distance covered will depend on the time; the greater the time, the greater will be the distance. This could be stated thus:

Distance depends on time,
or distance is proportional to time,
or distance $\propto$ time,
or distance :: time.

These statements show that the signs $\propto$ and :: mean 'is proportional to'. In this case, if the time is doubled, the distance will also be doubled, so one quantity is said to be in *direct proportion* to the other.

When an increase in one quantity causes an equal decrease in the other, the term *inverse proportion* is used. For example, this arises when speed is considered in relation to the time taken to cover a given distance.

So

$$\text{Speed} \propto \frac{1}{\text{time}} \text{ or speed} :: \frac{1}{\text{time}}$$

or speed is inversely proportional to time.

Time is written under a numerator of 1 when it is inversely proportional to the other quantity.

(The expression $\frac{1}{\text{time}}$ is called the *reciprocal* of time; in a similar way $\frac{1}{x}$ is the reciprocal of x.)

In respect to the example of speed and time, it will be realised that doubling the speed has the effect of halving the time taken to cover a certain distance.

Direct proportion

Example 1

If a car travels 135 km on 3 litres of petrol, how far will it travel on 7 litres?

On 3 litres car travels 135 km

On 1 litre car travels $\frac{135}{3} = 45$ km

On 7 litres car travels $45 \times 7 = 315$ km

This is called the *unitary method*: an alternative solution is as follows.

Distance $\propto$ quantity of petrol

so the 2 quantities are in direct proportion.

Stating the problem as two ratios,

ratio of distances :: ratio of petrol consumption

so $$135:x :: 3:7$$

The terms at each end, namely 135 and 7 are called the *extremes* and the middle terms, x and 3, are called the *means.*

The product of the *means* = the product of the *extremes*

so $$x \times 3 = 135 \times 7$$

and $$x = \frac{135 \times 7}{3} = 315 \text{ km}$$

Example 2

5 litres of oil costs £1·50. What will be the cost of 9 litres of this type of oil?

5 litres costs £1·50

$$\text{1 litre costs } \frac{£1{\cdot}50}{5}$$

$$\text{9 litres costs } \frac{£1{\cdot}50}{5} \times 9 = £0{\cdot}30 \times 9 = £2{\cdot}70$$

Alternative method

Quantity :: Cost

5 : 9 :: £1·50 : x

$$5x = 9 \times £1{\cdot}50$$

$$x = \frac{9}{5} \times £1{\cdot}50 = £2{\cdot}70$$

Inverse proportion

Example 1

A car travelling at a speed of 80 km/h takes 50 minutes to cover a given distance. How long will it take to travel a similar distance when the speed is 25 km/h?

Unitary method

Time taken at 80 km/h = 50 min

Time taken at 1 km/h = 50 x 80 (N.B. 80 times as long).

$$\text{Time taken at 25 km/h} = \frac{50 \times 80}{25} = 160 \text{ minutes.}$$

Alternative method

Time taken is inversely proportional to speed

$$\text{Time taken} :: \frac{1}{\text{speed}}$$

Note that both ratios are written in the same order — the 50 min and the 80 km/h are both stated first on each side of the :: sign.

$$\text{so } 50 : x :: \frac{1}{80} : \frac{1}{25}$$

$$50 \times \frac{1}{25} = x \times \frac{1}{80}$$

$$\frac{50}{25} = \frac{x}{80}$$

$25 \times x = 50 \times 80$ ← This is altered to a single line form by cross-multiplying. The numerator of one fraction is multiplied by the denominator of the other fraction

so $x = \frac{50 \times 80}{25}$

$x = 160$ minutes.

Exercises 1.4

1. Find the average of 406, 323, 564 and 731.
2. Find the average of 30·5, 52·7, 3·6 and 0·64.
3. Find the mean of 6, 4·2, 3·71, 7·49 and 5·25
4. On 26 litres of fuel a vehicle completes the following journeys: 62 km, 34 km, 85 km, 23 km and 32·6 km. What is the average petrol consumption in km/litre and litres/100 km?
5. A car travels 20 km at 80 km/h and 10 km at 40 km/h. What is its average speed in km/h?
6. The oil consumption of a vehicle was as follows:
   0·8 litre for a journey of 820 km
   0·3 litre for a journey of 240 km
   1·4 litre for a journey of 1280 km
   What is the average oil consumption for these three journeys?
7. The monthly sale by the Parts Department of radiator caps was as follows:
   12, 3, 8, 11, 19, 25, 27, 24, 12, 7, 18 and 2.
   What is the average monthly sale, (a) based on the yearly figures; (b) based on the first 6 months?
8. The sale of brake fluid was as follows:
   28·6 l for April, 42·5 l for May and 56·3 l for June
   What is the average sale per day assuming Sundays are included?

Express the following ratios in their lowest terms:

9. 28:49
10. 5·25 : 7·125
11. 1·2:0·96
12. 70 mm is to 2·8 m
13. 72 km/h is to 6 m/s
14. 18 seconds : 2 hours
15. A body jig is drawn to a scale such that the ratio between the drawing and the jig is 1: 5. What is the actual length of the jig if the corresponding distance on the drawing is 32 mm?
16. If a car travels 3·51 Mm on 3 litres of oil, how far will it travel on 2 litres?
17. A car, regularly maintained, is serviced 7 times in 42 Mm. At this rate when will the ninth service be required?
18. A road wheel makes 50 revolutions when the car moves 80 m. How many revolutions will the brake drum make when the vehicle travels 1 km?
19. An invoice shows that the total cost of 8 tyres is £56. What will be the charge for 15 tyres?

20. A bonus is to be shared between 3 groups of mechanics in the ratio 4:5:6. How much does each group receive if £90 is to be shared?

1.5 Percentages

A percentage is a fraction which has a denominator of 100, and serves as a useful common basis to compare fractional values, For example, 3 per cent is written as 3% and means $\frac{3}{100}$.

Conversion of a fraction to a percentage

Method — multiply the fraction by $\frac{100}{1}$

Example 1

Convert 0·375 to a percentage

$$0{\cdot}375 = 0{\cdot}375 \times \frac{100}{1} = 37{\cdot}5\%$$

Example 2

Convert 0·65 to a percentage.

$$0{\cdot}65 = 0{\cdot}65 \times \frac{100}{1} = 65\%$$

Conversion of a % to a fraction

Method — place the percentage over a denominator of 100 and reduce to its lowest form.

Example 3

Convert 23% to a decimal fraction

$$23\% = \frac{23}{100} = 0{\cdot}23$$

Example 4

Convert 22% to a decimal fraction.

$$22\% = \frac{22}{100} = 0{\cdot}22.$$

The following % are often used:

$$5\% = \frac{5}{100} = 0{\cdot}05 \text{ or } \frac{1}{20}$$

$$10\% = \frac{10}{100} = 0{\cdot}1 \text{ or } \frac{1}{10}$$

$$12\tfrac{1}{2}\% = \frac{12\frac{1}{2}}{100} = \frac{25}{200} = 0{\cdot}125 \text{ or } \frac{1}{8}$$

$$20\% = \frac{20}{100} = 0{\cdot}2 \text{ or } \frac{1}{5}$$

$$25\% = \frac{25}{100} = 0{\cdot}25 \text{ or } \frac{1}{4}$$

$$33\tfrac{1}{3}\% = \frac{33\frac{1}{3}}{100} = \frac{100}{300} = 0{\cdot}\dot{3} \text{ or } \frac{1}{3}$$

$$50\% = \frac{50}{100} = 0{\cdot}5 \text{ or } \frac{1}{2}$$

$$66\tfrac{2}{3}\% = \frac{66\frac{2}{3}}{100} = \frac{200}{300} = 0{\cdot}\dot{6} \text{ or } \frac{2}{3}$$

$$75\% = \frac{75}{100} = 0{\cdot}75 \text{ or } \frac{3}{4}$$

Application of %

Example 5

Find 20% of 2·25 litres

2·25 x 0·20

20% of 2·25 litres = 0·45 litre

Example 5

Due to increased costs the labour charge of a certain establishment is raised from £1·20 to £1·40. What is the percentage increase?

Increase in charge = 140 – 120 = 20p

$$\text{Proportional increase} = \frac{\text{Increase in charge}}{\text{Original charge}} = \frac{20}{120} = \frac{1}{6}$$

$$\%\text{ increase} = \frac{1}{6} \times \frac{100}{1} \triangleq 16{\cdot}7\%$$

Whenever % increases or decreases have to be calculated, it is necessary to remember that the variation is always based on the *original value*.

Example 6

Replacing the carburetter with a petrol injection system on a certain engine raises the brake power by 10%. What power will be developed by the new system if the original arrangement developed 60 kW?

Increase in power = 10% of 60
= 0·10 x 60 = 6 kW

So power developed by modified engine = 60 + 6
= 66 kW

Alternative method

Power developed by modified engine = 1·10 x 60
= 66 kW

In a similar way, if the power is decreased by 10% then new power output = 0·90 x 60 = 54 kW.

Exercises 1.5

Express as fractions:

1. 2·5% 2. 5·5% 3. 26% 4. 136%

Express as a %:

5. 0·6 6. 0·18 7. 1·28 8. 0·217

9. Find 6% of 3·125 litres
10. Find 20% of 0·7 metres
11. Find 33% of 0·15 m
12. Find 105% of 1·25 kg
13. In order to prevent damage to a cooling system, an antifreeze solution is mixed with water. How much antifreeze is necessary for a system of 6·76 l capacity if it is recommended that the quantity of anti-freeze is to be 25% of the cooling system's capacity?
14. How much water is added to a 12 litre cooling system if 20% of the system's capacity is anti-freeze solution?
15. A cooling system contains 2·5 litre of anti-freeze and 7·5 litre of water.
    (a) What is the % of anti-freeze in the cooling system and
    (b) what is the ratio of anti-freeze to water?
16. A test of a new type of tyre reveals that the life is increased by 4%. What was the distance expected from the original tyre if the new tyre lasts for 41·6 Mm.
17. It is decided that an account for £13·80 should be reduced by 10%. What is the value of the amended account?
18. The cost price of a component is £7·50. What is the retail charge if the buying price is 20% less than the selling price?
19. During the month of March, 31% of the firm's maximum stock of sparking plugs was sold. Assuming the maximum stock is 400, what was the average sale per day?
20. A job, for which the mechanic was allowed 12 h 30 min, was completed in 11 h 30 min. What percentage of the allotted time does the mechanic save?

1.6 Conversions— Imperial system to SI

It is expected that in your daily work you will be dealing with a wide range of vehicles – some new, some old. For many years this country used the Imperial system of units so a large number of the older vehicles will be dimensioned in this type of unit. Repairs of these vehicles means that the appropriate service manual is sometimes consulted to check a particular clearance or size of component. Also it is likely that occasions will arise when an instrument calibrated in SI units is used to measure a component dimensioned in the Imperial system. In order to cover these situations this chapter has been included.

Most conversions will be performed by using various aids recommended by BSI, but in order to give an appreciation of the relationship between the size of unit the following values are given. Except where stated the values are approximated to suit most applications dealing with repairs to vehicles. Extra special care must be exercised where safety is concerned, e.g. lifting equipment, since the SI primary primary unit is larger than that which was used in the Imperial system.

Length

1 yard = exactly 0·914 4 metre
1 inch (in) $\triangleq$ 25·4 mm
1 ft (ft) $\triangleq$ 304·8 mm
1 mile $\triangleq$ 1·6 km

Mass

1 pound (lb) = exactly 0·453 592 37 kg
1 ounce (oz) $\triangleq$ 28·35 g
1 ton $\triangleq$ 1016 kg = 1·016 tonne

Capacity

1 pint (pt) $\triangleq$ 0·568 litre
1 gallon (gal) $\triangleq$ 4·536 litre

Other conversions are shown in the Science section of the book.

To convert from Imperial to SI, the Imperial value is multiplied by the appropriate SI 'equivalent'.

Example 1

Convert 2·5 inches to SI

There are 25·4 mm in 1 inch

2·5 inches = 2·5 x 25·4
= 63·5 mm

Example 2

Convert 30 miles to SI

There are 1·6 km in 1 mile

30 miles = 30 x 1·6
= 48 km

Example 3

Convert 0·002 inch to SI

There are 25·4 mm in 1 inch

0·002 inch = 0·002 x 25·4
= 0·0508 mm
= 50·8 μm

Example 4

Convert 2500 lb to SI

There are 0·454 kg in 1 pound

2500 lb = 2500 x 0·454
= 1135 kg
= 1·135 tonne

Example 5

Convert 3 pints to SI

There are 0·568 litres in 1 pint

$$3 \text{ pints} = 3 \times 0{\cdot}568 = 1{\cdot}704 \text{ litres}$$

Example 6

Convert 35 miles/gallon to SI.

There are 1·6 km in a mile and 4·536 litres in a gallon

$$\begin{aligned} 35 \text{ miles/gallon} &= 35 \times 1{\cdot}6 \\ &= 56 \text{ km/gallon} \\ &= 56 \text{ km}/4{\cdot}536 \text{ litres} = \frac{56}{4{\cdot}536} \\ &\triangleq 10{\cdot}4 \text{ km/litre} \end{aligned}$$

OR

$$\begin{aligned} 10{\cdot}4 \text{ km/litre} &= \frac{100}{10{\cdot}4} \text{ litres/100 km} \\ &\triangleq 9{\cdot}61 \text{ litres/100 km} \end{aligned}$$

Exercises 1.6

Convert, with the aid of conversion tables if available, the following:

1. 1·7 inches to mm
2. 25 miles to km
3. 0·022 inches to mm
4. 1·5 tons to tonnes
5. 500 lb to kg
6. 22 oz to g
7. 7·5 gallons to litres
8. 8 pints to litres
9. 30 mile/h to km/h
10. 15 mile/gallon to km/litre
11. A new crankshaft has a diameter of 2·375 inches. How much undersize in mm is a journal which measures 59·325 mm.
12. The bore of an unworn, or standard size, cylinder is 88·9 mm. What will be the diameter in mm after it has been rebored 0·010 inch oversize?
13. The clearance between a certain shaft and bearing should be 0·002 inch. What will be the diameter of a new shaft if it is to fit a bearing of diameter 50 mm?
14. During a test on a valve spring a mass of 80 lb is suspended on a spring to deflect it to its working length of 0·3125 in. What are the metric equivalents of these values?
15. Operating at a speed of 40 mile/h a car uses fuel at the rate of 2 gallons/h. At this speed and consumption rate, what distance in km will it travel on 1 gallon of fuel?
16. A speedometer, which indicates a speed 10% higher than the actual value, shows a reading of 55 mile/h. How long will it take to travel 16 km if this speed is maintained?
17. A car travels 87·5 miles on 3·5 gallons of petrol. What distance in km will it travel on 8 litres of petrol?

18. A lifting hoist is capable of lifting a mass of 3 tons. What is the SI equivalent of this value.
19. One turn of a road wheel moves a vehicle forward to the extent of 88 inches. Calculate the number of revolutions made by the wheel when the vehicle travels a distance of 4·47 km.
20. The drain tap of an engine cooling system leaks at the rate of 40 ml per hour. Assuming the leakage continues at this rate, what will be the time taken for 2 pints to drain from the system?

1.7 Areas and volumes

During your course you will discover many instances where the area or volume of a figure has to be determined. It arises in the calculation of engine power, compression ratio, capacity of tanks and many other instances. Only the basic shapes are considered in this chapter, but knowledge of the methods used enables more intricate shapes to be handled.

Perimeter and area

The perimeter of a figure is the distance around the outside; the area represents the extent of the surface.

Square

Perimeter. The length of a square is equal to the breadth, so if l is the length of one side then:

$$\text{Perimeter} = 4 \times l$$

When $l = 3$ m, the perimeter is

$$4 \times 3 = 12 \text{ metres}$$

Area. Consider a square having a length of side of 3 metres. If this is drawn to scale and divided up as shown in the diagram, it will form nine smaller squares. Each small square will have a side length of 1 ft and would represent 1 square metre (Written as 1 m^2).

So

Area of this square = 3×3
(3 layers of 3)
= 9 m^2
Area of any square = $l \times l$
= l^2

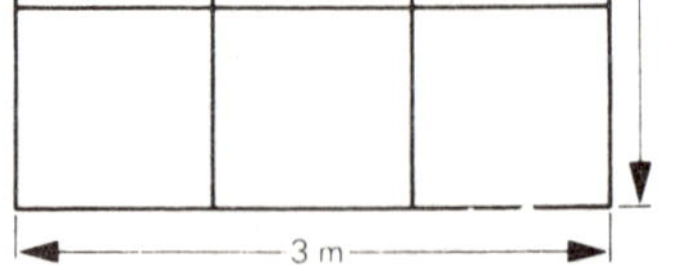

(This is a short way of writing $l \times l$ — it is read as 'l squared'.)

Sometimes it is necessary to use a smaller unit than the square metre (m^2). If a square metre is divided up into square centimetres, it will be seen that:

$1\ m^2 = 100 \times 100$
$= 10\,000\ cm^2$

The area of a square having a side of 3 m would be 9 m^2 or $9 \times 10\,000\ cm^2 = 90\,000\ cm^2$

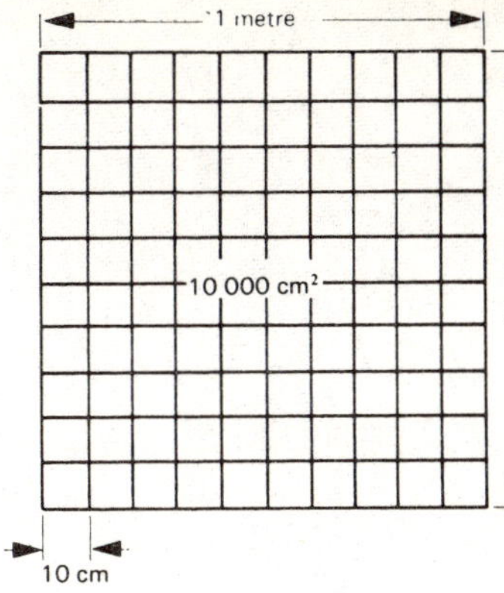

This example shows:

$100\ mm^2 = 1\ cm^2$
$100\ cm^2 = 1\ dm^2$
$100\ dm^2 = 1\ m^2$

Rectangle

Perimeter = $2l \times 2b$

Area = length x breadth
or Area = $l \times b$

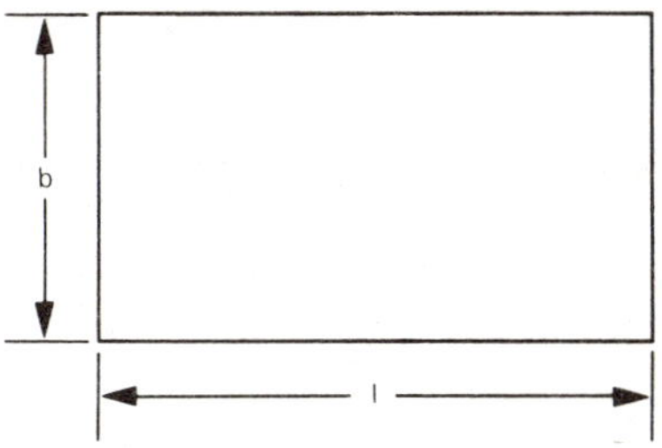

For a rectangle 5 m x 2 m

perimeter = $2l \times 2b$
$= 2 \times 5 + 2 \times 2$
$= 10 + 4 = 14$ m
area $= l \times b$
$= 5 \times 2 = 10\ m^2$

Triangle

The diagram shows a right-angled triangle — i.e. a triangle having one angle of 90°. It will be seen that side 'c' is the longest — this side is called the *hypotenuse.*

Perimeter = a + b + c.
Area. The triangle may be considered as half a rectangle having sides of length 'a' and 'b'.

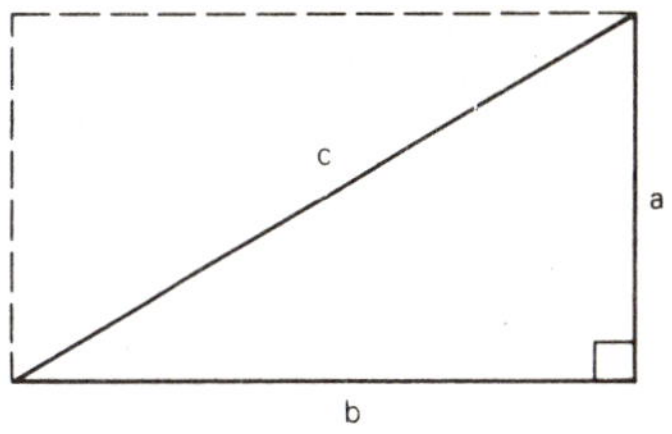

Area of triangle $= \dfrac{\textit{Area of rectangle}}{2}$

$= \dfrac{a \times b}{2}$

$= \frac{1}{2} a \times b$

or area $= \frac{1}{2}$ base x perpendicular height.

For a triangle having side lengths of 3 m, 4 m and 5 m,

perimeter = 3 + 4 + 5
= 12 m
area $= \frac{1}{2}$ base x perpendicular height
$= \frac{1}{2} \times 4 \times 3$
$= 6\ m^2$

(If the ratio between the sides of a triangle is 3:4:5, then the angle opposite to the *hypotenuse* is 90°. In the above example the 90° angle ensures that the perpendicular height is 3 m.)

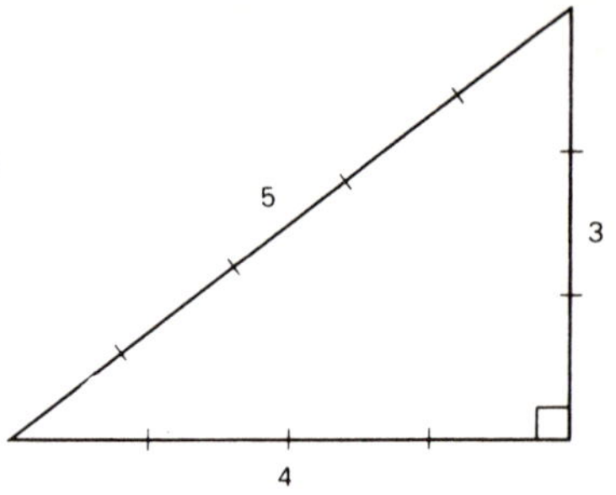

Circle

Circumference is the perimeter of the circle.
Radius is the distance from the centre to the circumference.
Diameter is the distance, measured through the centre, from circumference to circumference.

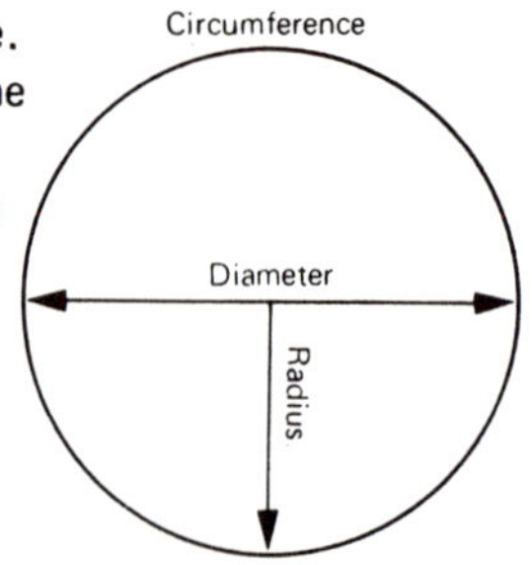

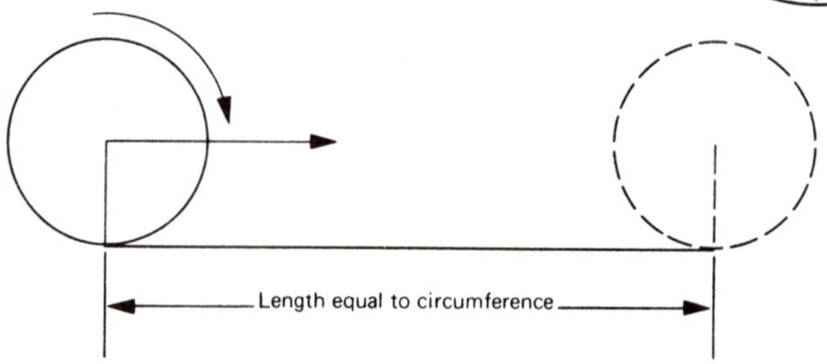

Perimeter. The perimeter of any circular object can be determined by rolling it through one complete revolution along a flat surface. Measuring this distance and comparing it with the diameter will show that the ratio $\frac{\text{circumference}}{\text{diameter}}$ is always constant irrespective of the size of the circle. Represented by the Greek letter π (pronounced pi), this ratio is 3·142 correct to four significant figures or approximately $3\frac{1}{7}$ which can also be written as $\frac{22}{7}$.

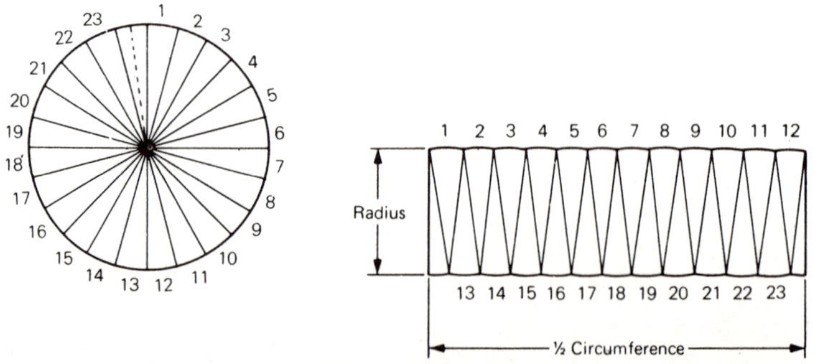

So:

$$\text{perimeter of circle} = \text{circumference} = \pi d = 2\pi r$$

where r = radius

d = diameter

(N.B. When symbols are placed adjacent to each other it means that the symbols are multiplied together.)

Area. Dividing the circle up into an infinite number of triangles and arranging these into a rectangular form shows that the

$$\begin{aligned}\text{area of a circle} &= \text{length} \times \text{breadth of equivalent rectangle}\\ &= \tfrac{1}{2}\,\text{circumference} \times r\\ &= \tfrac{1}{2} \times 2\pi r \times r \text{ since circumference} = 2\pi r\\ &= \pi r \times r\\ &= \pi r^2\end{aligned}$$

Since $r = \dfrac{d}{2}$ then:

$$\text{area} = \pi \times \frac{d}{2} \times \frac{d}{2} = \frac{\pi d^2}{4}$$

Example 1

What is the perimeter and area of a circle of diameter 70 mm?

Perimeter = circumference = $\pi d = \frac{22}{7} \times 70 = 220$ mm

$$\begin{aligned}\text{Area} &= \pi r^2\\ &= \frac{22 \times 35 \times 35}{7}\\ &= 3850\ \text{mm}^2 \quad \text{or} \quad 38{\cdot}5\ \text{cm}^2\end{aligned}$$

or

$$\begin{aligned}\text{Area} &= \frac{\pi d^2}{4}\\ &= \frac{\overset{11}{\cancel{22}} \times \cancel{7} \times 7}{\cancel{7} \times \underset{2}{\cancel{4}}} = \frac{77}{2} = 38{\cdot}5\ \text{cm}^2.\end{aligned}$$

Volume

The volume, or cubic content, of a rectangular solid is obtained by the expression

$$\text{volume} = \text{area of end} \times \text{length}$$

This statement may be verified by considering the rectangular figure shown here. Each small cube represents a cubic centimetre; in this case the area of one end shows the presence of four cubes and the depth of the solid shows that there are four layers. Altogether there are

$2 \times 2 \times 4 = 16$ cubes of 1 cubic centimetre

So:

volume of a rectangular solid

$$\begin{aligned}&= \text{length} \times \text{breadth} \times \text{height}\\ &= l \times b \times h\end{aligned}$$

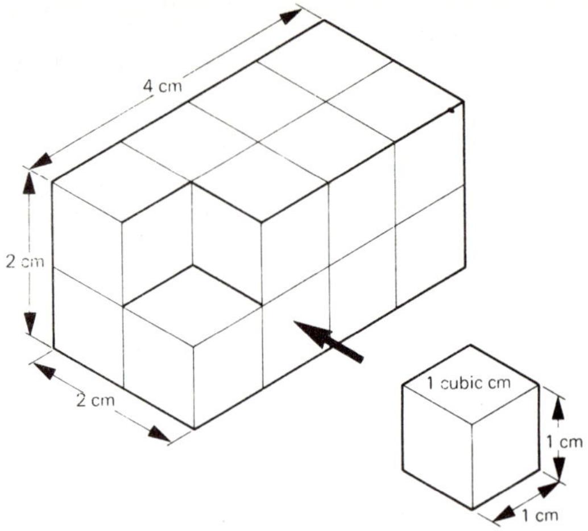

A cubic cm can be divided up into smaller cubes of 1 cubic millimetre. The diagram shows that there are 10 layers of 100. So:

$$1 \text{ cm}^3 = 1000 \text{ mm}^3$$

Whenever a volume is calculated the units must be the same throughout; i.e. if '*l*' is in cm, then '*b*' and '*h*' must also be in cm.

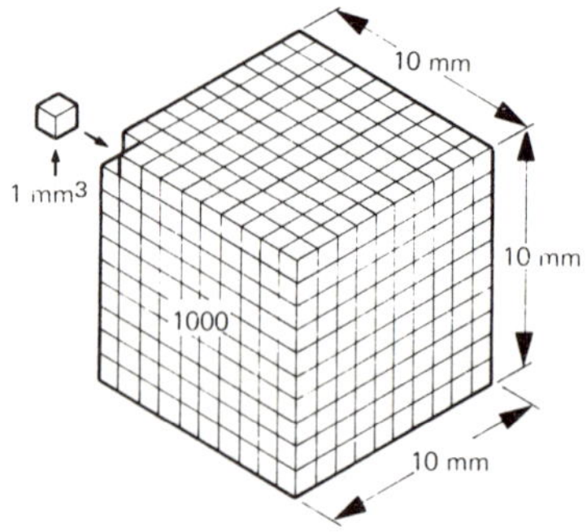

This example shows:

$$1000 \text{ mm}^3 = 1 \text{ cm}^3$$
$$1000 \text{ cm}^3 = 1 \text{ dm}^3$$
$$1000 \text{ dm}^3 = 1 \text{ m}^3$$

Cylinders may be treated in a similar manner to rectangular solids.

Volume = area of end x l

$= \pi r^2 \times l$

$= \pi r^2 l$ or $\dfrac{\pi d^2 l}{4}$

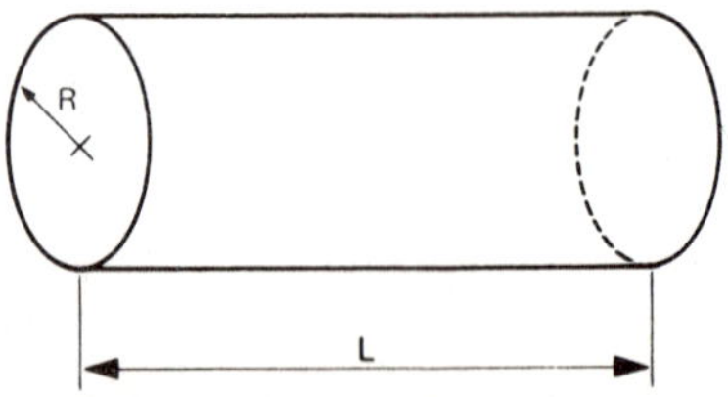

These expressions can only be used for solids having a regular cross-section; i.e. a constant area throughout the length.

Example 2

A tank having internal dimensions of 0·8 m x 0·2 m x 0·5 m is completely filled with petrol. What is:

a) the volume of the tank; b) the capacity of the tank?

Volume = length x breadth x height
= 0·8 x 0·2 x 0·5
= 0·08 m^3

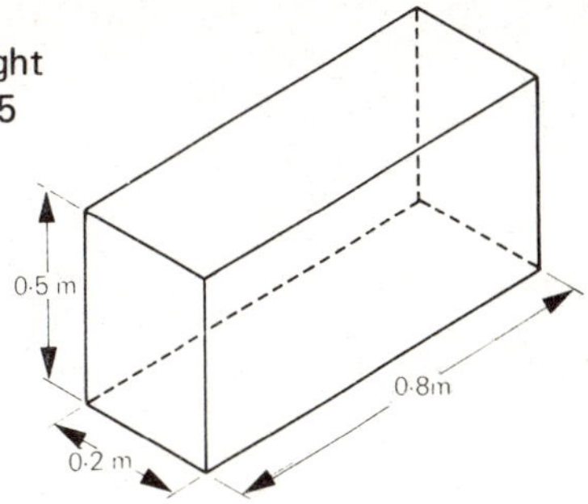

'Capacity' indicates the maximum quantity of petrol that the tank will contain. This is often expressed in litres and it is useful to remember that 1 dm^3 (0·001 m^3) is equivalent to 1 litre. So

$$\text{Capacity of this tank} = \frac{\text{volume in dm}^3}{1\text{ dm}^3}\text{ litres}$$

$$= \frac{80}{1}$$

$$= 80\text{ litres}$$

Problems which require the capacity in litres are simplified if the dimensions are changed to decimetres (0·1 metre). Example 2 could be rewritten as:

$$\text{Volume} = l \times b \times h$$
$$= 8 \times 2 \times 5\text{ dm}^3$$
$$= 80\text{ litres} \quad (\text{since } 1\text{ dm}^3 = 1\text{ litre})$$

The reader is advised to use the following sequence when dealing with problems:

1. Write down the appropriate formula.
2. Transpose if necessary.
3. Select units which give suitable numbers even if the unit is non-standard.
4. Substitute the values in the formula and perform the numerical operation.
5. Arrange the value to give the answer in a standard unit.

Example 3

A cylindrical tank of diameter 20 cm and length 35 cm is filled with water. What is the mass of the water contained in this tank given that 1 cm^3 of water has a mass of 1 gramme.

$$\text{Volume} = \frac{\pi d^2}{4} \times l$$

In this example it is an advantage to work in cm

$$\text{Volume} = \frac{22 \times \overset{100}{\cancel{400}} \times \overset{5}{\cancel{35}}}{\cancel{7} \times \cancel{4}}$$

$$= 110 \times 100$$

$$= 11\,000 \text{ cm}^3 = \text{capacity}$$

Mass of water = 11 000 g = 11 kg.

Example 4

Determine the maximum mass of petrol of relative density 0·72 that could be contained in a tank having a regular cross-sectional area of 400 cm^2 and length 30 cm.

$$\begin{aligned}\text{Volume of tank} &= \text{area of end} \times \text{length}\\ &= 400 \times 30\\ &= 12\,000 \text{ cm}^3 \quad (\text{or } 12 \times 10^3 \text{ cubic centimetres})\end{aligned}$$

The relative density of a substance is the ratio of the mass of a given volume of the substance to the mass of an equal volume of water.

$$\text{Relative density} = \frac{\text{mass of any volume of the substance}}{\text{mass of an equal volume of water}}$$

In this case the mass of the petrol is 0·72 of the mass of an equal volume of water.

Since volume = 12 000 cm^3

Mass of water that could be contained in the tank = 12 000 grammes = 12 kg

$$\begin{aligned}\text{Mass of petrol in the tank} &= 12 \times \text{relative density}\\ &= 12 \times 0{\cdot}72\\ &= 8{\cdot}64 \text{ kg}\end{aligned}$$

Example 5

Find the swept volume of an engine cylinder having a bore of 76 mm and a stroke of 77 mm

$$\text{Swept volume of cylinder} = \frac{\pi d^2 \times l}{4}$$

$$= \frac{22 \times 76 \times \overset{19}{\cancel{76}} \times \overset{11}{\cancel{77}}}{\cancel{7} \times \cancel{4}} \text{ mm}^3$$

$$= 349\,448 \text{ mm}^3 \text{ or } 349{\cdot}448 \text{ cm}^3$$

$$\simeq 350 \text{ cm}^3$$

This value could be used to find the compression ratio (see Chapter 2 F of M.V.T.).

If the clearance volume is 50 cm^3 then:

$$\text{Compression ratio} = \frac{\text{swept vol} + \text{clearance vol}}{\text{clearance vol}}$$

$$= \frac{350 + 50}{50}$$

$$= \frac{400}{50}$$

$$= 8:1$$

N.B. The 50 must not be cancelled – the addition must be completed first.

Summary

Perimeter

Square	4 x length of 1 side	or $4l$
Rectangle	2 x length + 2 x breadth	or $2l + 2b$
Triangle	Total length of the three sides	or a + b + c
Circle	Circumference	or $2\pi r$ or πd

Area

Square	length x breadth	or l^2
Rectangle	length x breadth	or $l \times b$
Triangle	$\frac{1}{2}$ base x perpendicular height	or $\frac{1}{2}b \times h$
Circle	πr^2 or $\frac{\pi d^2}{4}$	

Volume

Volume of any rectangular solid = area of end x height = $l \times b \times h$

Cylinder = area of end x length = $\pi r^2 l$ or $\frac{\pi d^2 l}{4}$

1 cubic decimetre = 1 litre
1 cm^3 of water = 1 gramme

$$\text{Relative density} = \frac{\text{mass of any volume of a substance}}{\text{mass of an equal volume of water}}$$

(In the past the expression specific gravity was used instead of relative density.)

$$\text{Compression ratio} = \frac{\text{swept volume + clearance volume}}{\text{clearance volume}}$$

Exercises 1.7

1. A rectangular windscreen has a length of 1300 mm and a breadth of 350 mm.
   (a) Find the area of the screen; (b) Determine the length of rubber strip required to seal the windscreen.
2. A road wheel having an effective diameter of 700 mm turns through 1500 revolutions. What distance will the vehicle travel during this period?
3. How many turns are made by a wheel of diameter 700 mm to cover the distance of 11 km?
4. Carbon of thickness 1 mm covers the flat crown of a piston of diameter 84 mm. What is the volume of this deposit?
5. A piece of metal, cut to form a triangle, has a base of 105 mm and a perpendicular height of 52 mm. What is the area of one face?

6. A rectangular petrol tank has internal dimensions 1 m x 0·5 m x 0·25 m. What quantity, in litres of petrol can be delivered into the tank if it already holds 0·4 of its total capacity of petrol?
7. A rectangular sump 250 mm x 200 mm is filled with oil to a depth of 75 mm. Find the weight of this oil if 1 litre of oil has a mass of 0·9 kg.
8. An engine has a bore of 80 mm and a stroke of 84 mm. Calculate the swept volume of one cylinder.
9. (a) What is the circumference of an exhaust pipe having an external diameter of 44 mm? (Answer to the nearest mm and take π as $\frac{22}{7}$.)
   (b) If the wall thickness of this pipe is 1 mm, calculate the internal cross-sectional area of the pipe.
10. A cubical tank has a side length of 0·3 m. What is the:
    (a) total surface area
    (b) capacity in litres?

1.8 Simple equations and formulae

Use of symbols

In the previous chapter there were many instances where letters were used to represent numbers; e.g. the area of a rectangle is $l \times b$. The letters are called symbols and these are used in preference to writing the terms in full.

When using symbols it is important to state the meaning of each together with the unit of measurement. British Standards Institute (B.S.I.) recommend the use of certain symbols and these should be adopted whenever possible. If the meaning of the symbols is not universally understood then the relationship may be laid out as follows:

$$a = l \times b \quad \text{where } a = \text{area of rectangle (m}^2\text{)}$$
$$l = \text{length (m)}$$
$$b = \text{breadth (m)}$$

or shorter still to

$$a = lb$$

since it is accepted that when the symbols are placed next to each other it means that the symbols are multiplied together.

Formulae

The various expressions used in this chapter show the relationship between stated quantities; e.g. the expression $a = \pi r^2$ indicated the 'connection' between the radius and area of a circle. This general arrangement of symbols is called a *formula.* Having discovered the appropriate formula it is possible to insert the respective numbers and calculate the result.

Example 1

The compression ratio of an engine is given by the formula:

$$R = \frac{V_s + V_c}{V_c} \quad \text{where } R = \text{compression ratio}$$
$$V_s = \text{swept volume cm}^3$$
$$V_c = \text{clearance volume cm}^3$$

An engine having a swept volume of 400 cm^3 and clearance volume of 50 cm^3 would have a compression ratio of:

$$R = \frac{400 + 50}{50} = \frac{450}{50} = 9:1$$

Example 2

The length of a rod after heating is given by the formula:

$L = lt\alpha + l$ where l = original length (m)
t = temperature rise (°C)
α = coefficient of expansion
L = final length (m)

Determine the final length if l = 10 m, t = 50°C and α = 0·00002.

Substituting values,

$$\begin{aligned} L &= 10 \times 50 \times 0{\cdot}000\,02 + 10 \\ &= 500 \times 0{\cdot}000\,02 + 10 \\ &= 5 \times 0{\cdot}002 + 10 \\ &= 0{\cdot}01 + 10 \\ L &= 10{\cdot}01 \text{ m} \end{aligned}$$

(You will notice that the 10 is not added until the multiplication has been completed. In a problem where various signs are employed the order of priority 'B D M A S' is used. This stands for brackets, division multiplication, addition and subtraction.)

When symbols are replaced by figures it will be seen that the formula can be worked out providing only one unknown value exists.

Where two formulae use the same symbol, e.g. $a = lb$; $a = \pi r^2$; this may cause confusion especially if they were both used in the same problem. In this case the symbol for area may be separated by using forms such as A and a or a_r and a_c. Here a_r and a_c represent the area of a rectangle and area of a circle respectively.

In a similar manner the form of symbol can act as a guide to the unit; e.g. there are occasions where 'length' is quoted in metres and other cases where it is stated in mm. These can be easily recognised if L and l are used as the major and minor unit respectively.

$$\therefore\ L = \frac{l}{1000}$$ where L = length (metres)
l = " (millimetres)

Example 3

A cylindrical, hydraulic brake reservoir is replaced by a new container which is the same length but twice the diameter of the original. How will this alteration affect the capacity?

Capacity $\propto$ area

Let a = original cross-sectional area
A = final cross-sectional area

$$\therefore\ a = \pi r^2 \quad \text{and} \quad A = \pi R^2$$

since $R = 2r$

then $A = \pi(2r)^2$ if $2r$ is substituted for R.

Multiplying $2r$ by itself gives $4r^2$

$\therefore$ $A = 4\pi r^2$

Now $a = \pi r^2$

So $A = 4 \times a$

i.e. When the diameter or radius is doubled, the area is increased by four times the original.

Answer. The capacity will be four times as great.

Addition and subtraction of symbols

A symbol represents a number, so it is possible to add or subtract symbols.

Since $5 + 5 + 5 + 5 = 4 \times 5$
then $a + a + a + a = 4 \times a = 4a$

In a similar manner:

$5 + 5 + 3 + 3 + 3 + 3 = 2 \times 5 + 4 \times 3$
and $a + a + b + b + b + b = 2a + 4b$

Since 'a' and 'b' represent different numbers, the expression 2a + 4b cannot be reduced further until the values of 'a' and 'b' are known.

Addition and subtraction must be restricted to *like terms*; i.e. the 'a's must be kept separate from the 'b's.

Example 1

Simplify 6a + 4b + 3c + 2a + 3b + c.

Grouping 'like' terms under each other and adding,

$$\begin{array}{r} 6a + 4b + 3c \\ 2a + 3b + c \\ \hline 8a + 7b + 4c \end{array}$$

The number is placed in front of the symbol; a value of 1 is implied when the symbol has no number. When a sign is not written in front of the symbol it is taken as +, so 8a means + 8a.

Example 2

Add 5a + 6b to 3a − 2b.

$$\begin{array}{r} 5a + 6b \\ 3a - 2b \\ \hline 8a + 4b \end{array}$$

Adding +6b to −2b gives +4b. The meaning of this can be seen if you imagine that you have £6 in your pocket but owe £2. Your financial position is + 6 − 2 = £4.

Example 3

Subtract 4b + 2c from 3c + 6b.

Arranging the terms as in normal subtraction, but keep the like terms under each other.

$$\begin{array}{r} 6b + 3c \\ 4b + 2c - \\ \hline 2b + \;\; c \end{array}$$

Example 4

Subtract $-3a + 2b$ from $5a - 6b$

$$\begin{array}{r} 5a - 6b \\ -3a + 2b \\ \hline 8a - 8b \end{array}$$

The simple method of dealing with minus signs on the bottom line is to mentally change the signs on this line and proceed as in addition.

The signs in the question would become

$$\begin{array}{r} 5a - 6b \\ 3a - 2b \\ \hline 8a - 8b \end{array}$$

Multiplication and division

When symbols have to be multiplied or divided, the 'rule of signs' must be applied. This rule states:

like signs give +
unlike signs give −
or $+ \times + = +$
$- \times - = +$
$+ \times - = -$
$- \times + = -$

so using this rule:

$$a \times a = +a^2$$
$$-a \times a = -a^2$$

Example 1

Multiply 3a by 5a.

Both signs are + so the answer will be +

$$\therefore \; 3a \times 5a = 15a^2$$

The numbers 3 and 5 placed in front of the symbol are called the *coefficients* and the 2 is called the *index* (plural, indices). Writing $3a \times 5a$ in another form it becomes:

$$3 \times 5 \times a \times a = 15a^2$$

To multiply one term by another, the coefficients are multiplied together and the indices of similar terms are added.

Example 2

$6a^2 \times 5a^3 = 30a^5$

Example 3

$4a^2b(-2a) = 4a^2b \times (-2a) = -8a^3b$

Example 4

$-3a^2b^4c \times 4a^3b^5c^6 = -12a^5b^9c^7$

When dividing one quantity into another:

a) apply the rule of signs,
b) divide the coefficients and
c) subtract the indices of similar terms.

Example 5

$$7a^2 \div 4a = \frac{7a^2}{4a} = 1{\cdot}75a \qquad \text{Check. } 1{\cdot}75a \times 4a = 7a^2$$

Example 6

$$-12a^5 \div 4a^2 = -\frac{12a^5}{4a^2} = -3a^3 \qquad \text{Check. } -3a^3 \times 4a^2 = -12a^5$$

Example 7

$$-20a^2b^5c^4 \div (-5ab^3c^3) = \frac{-20a^2b^5c^4}{-5ab^3c^3} = 4ab^2c$$

$$\text{Check. } 4ab^2c \times (-5ab^3c^3) = -20a^2b^5c^4$$

Brackets

In the previous section it was seen that a bracket was used to separate the signs. In other cases the bracket can group quantities together. For example. Consider the perimeter of a rectangle of length l m and breadth b m.

The length of two sides would be $l + b$ and the perimeter would be $2(l + b)$.

If $l = 5$ m and $b = 3$ m then

$$\begin{aligned} \text{perimeter} &= 2(l + b) \\ &= 2l + 2b \\ &= 10 + 6 \\ &= 16 \text{ metres} \end{aligned}$$

In this case a multiplication sign is assumed to precede the bracket, so everything inside the bracket is multiplied by the 2. Generally it is *advisable* to deal with the figures inside the bracket first.

Brackets preceded by a plus sign can be removed without altering the signs so:

$$3 + (4 + 7 - 2) = 3 + 4 + 7 - 2$$

but a minus sign before the bracket means that all signs inside the bracket are changed:

$$3 - (4 + 7 - 2) = 3 - 4 - 7 + 2$$

Note that the sign before the 4 is taken as +.

Simple equations

Whenever values are substituted for symbols in a formula, an expression is produced which is called an *equation.* This is actually a balance of numbers which generally contains one unknown quantity.

Any symbol may be used for this unknown, but often the latter part of the alphabet is utilised. An equation can be formed if certain facts are known, as this example shows.

Example 1

When 3 litres of oil are drained from a sump, the remaining quantity corresponds to the half-full position. How many litres of oil did the sump contain if the total capacity is 12 litres?

Let x = quantity of oil in sump before oil was drained. Then

$$x - 3 = \tfrac{1}{2} \times 12$$

Having formed an equation we must now solve it; i.e. find the value of x. Since the equation represents a balance such that the left-hand side balances the right-hand side, it follows that whatever is done to one side must be done to the other side also.

This will be apparent in the solution of this and other examples. Applying this rule,

$$x - 3 = \tfrac{1}{2} \times 12$$
$$x - 3 = 6$$
$$x - 3 + 3 = 6 + 3 \qquad \text{Adding 3 to both sides.}$$
$$x = 9$$

∴ Quantity of oil contained in the sump was 9 litres.

Example 2

Solve $3x + 5 = 14$

$$3x + 5 - 5 = 15 - 5 \qquad \text{Subtracting 5 from both sides.}$$
$$3x = 9$$
$$\frac{3x}{3} = \frac{9}{3} \qquad \text{Dividing both sides by 3.}$$
$$x = 3$$

Example 3

Solve $7x - 4 = 3x + 6$

$$7x - 4 - 3x = 3x + 6 - 3x \qquad \text{Subtracting } 3x \text{ from both sides.}$$
$$4x - 4 = 6$$
$$4x = 6 + 4 \qquad \text{Adding 4 to both sides.}$$
$$4x = 10$$
$$x = \frac{10}{4} \qquad \text{Dividing by 4.}$$
$$x = 2{\cdot}5$$

Example 4

Solve $\dfrac{3x - 3}{2} = \dfrac{4x + 3}{5}$

The L.C.M. of 2 and 5 is 10
Multiplying both sides by 10

$$10\left(\frac{3x-3}{2}\right) = 10\left(\frac{4x+3}{5}\right)$$

$5(3x - 3) = 2(4x + 3)$

$15x - 15 = 8x + 6$

$7x - 15 = 6$ — Subtracting $8x$ from each side.

$7x = 6 + 15$ — Adding 15 to each side.

$7x = 21$

$x = 3$

The second line in the solution shows that the original equation may be simplified to a single line expression by *cross multiplying* – the first numerator is multiplied by the denominator of the second term and vice versa.

To simplify $\frac{2x}{7} = \frac{3}{2}$ cross

cross-multiply

$$\frac{2x}{7} \times \frac{3}{2}$$

$2x \times 2 = 7 \times 3$

$4x = 21$

$x = 5{\cdot}25$

Transposition of formulae

In a formula $A = \frac{1}{2}bh$ for the area of a triangle, it has been shown that A can be found by substituting values for b and h. Often the subject of the formula, which in this case is A, is given and another symbol such as b has to be found. This means that the formula must be rearranged to make it suitable for the calculation of b.

Example 1

Make b the subject in the formula

$$A = \tfrac{1}{2}bh$$

Remembering the rule for equations, b is transpositioned as follows:

$\frac{A}{h} = \frac{1}{2}b$ — Dividing both sides by h.

$\frac{2A}{h} = b$ — Multiplying both sides by 2.

$\therefore \quad b = \frac{2A}{h}$

This example may be considered in another manner since the solution shows that as the symbol disappears from one side it reappears on the other side – the only difference being in the sign. This transposition of symbols may be carried out if the following rules are applied:

$\times \overset{\curvearrowright}{=} \div$

$\div \overset{\curvearrowright}{=} \times$

$+ \overset{\curvearrowright}{=} -$

$- \overset{\curvearrowright}{=} +$

Taking the first rule as an example, when the symbol multiplies on one side, then on transferring the quantity to the other side of the equal sign, the sign is changed to ÷.

Example 2

Make b the subject in:

$$A = l \times b$$

Since l multiplies, then on transferring it to the other side the sign is changed to ÷.

$$A \div l = b$$

$$\text{or } b = \frac{A}{l}$$

Example 3

Make h the subject in:

$$A = \pi r^2 h$$

$$\therefore \frac{A}{\pi r^2} = h$$

h is multiplied by πr^2, so when πr^2 is transferred the sign is changed.

Example 4

Make t the subject in:

$$L - l = ltE$$

$$\frac{L - l}{lE} = t$$

Example 5

Make l the subject in:

$$V = \frac{\pi d^2 l}{4}$$

This may be written as:

$$\frac{V}{1} = \frac{\pi d^2 l}{4}$$

Cross-multiply to get the expression in a single line:

$$\pi d^2 = 4V$$

Since the left-hand side terms are all multiplied together, then

$$l = \frac{4V}{\pi d^2}$$

Example 6

Make 'd' the subject in:

$$A = \frac{\pi d^2}{4}$$

$$\pi d^2 = 4A \quad \text{by cross-multiplying}$$

$$d^2 = \frac{4A}{\pi}$$

Since '*d*' is squared (i.e. *d* x *d*), then to remove the square, the square root of both sides is taken. This becomes:

$$\sqrt{d^2} = \sqrt{\frac{4A}{\pi}}$$

or

$$d = \sqrt{\frac{4A}{\pi}}$$

This example shows that if the square is removed from one side, then it appears as a root on the other side.

Exercises 1.8

1. To convert a temperature from the Fahrenheit scale to the Celsius scale the following formula is often used:
   $C = \frac{5}{9}(F - 32)$ where F and C represent the temperature in degrees Fahrenheit and Celsius respectively.
   Using this formula convert a temperature of 95°F to °C.
2. The relationship between the electrical resistance, voltage and current is given by the expression:
   $R = \frac{E}{I}$ where R = resistance (ohm)
   E = voltage
   I = current (ampere)
   Calculate the resistance of a circuit which passes a current of 1·2 ampere when the voltage is 11·4 volts.
3. To determine the correct spindle speed for a drill cutting a material, the formula is:
   $S = \frac{\pi dN}{1000}$ where S = recommended cutting speed (m/min)
   d = diameter of drill (mm)
   N = spindle speed (rev/min)
   Determine the spindle speed for a drill of diameter 12 mm to cut steel at a speed of 36 m/min. Take π as $\frac{22}{7}$.
4. The compression ratio of an engine is given by:
   $R = \frac{V_s + V_c}{V_c}$ where R = compression ratio
   V_s = swept volume (cm³)
   V_c = clearance volume (cm³)
   Find the swept volume of an engine having a compression ratio of 8 to 1 and a clearance volume of 50 cm³.
5. The brake efficiency is given by the expression:
   $E = \frac{0{\cdot}39V^2}{S}$ where E = efficiency %
   V = speed km/h
   S = stopping distance, metres
   Calculate the stopping distance from a speed of 80 km/h if the brakes have an efficiency of 78%.
6. Add 6b + 4c to 7b − 5c.
7. Add $3x - 2y$ to $8y - 5x$.
8. Subtract 3a + rb from 5a + 3b.
9. Subtract $5x - 9y$ from $3x + 5y$.
10. Multiply $4a^2$ by $7a^3$.

11. Multiply $-7x^4y$ by $3x^2y^2$.
12. Divide $28a^4$ by $7a$.
13. Divide $42x^3y^5z$ by $7x^2y^3z$.
14. Simplify $15-(5+3+4)$.
15. Solve $3x-3=2x+{}_2$.
16. Solve $\dfrac{7x-5}{2}=\dfrac{3x+31}{5}$

Transpose the following formula:

17. $L=lt\alpha+l$ to make α the subject.
18. $V=\dfrac{\pi d^2h}{4}$ to make h the subject.
19. $V=\dfrac{\pi d^2h}{4}$ to make d the subject.
20. $E=IR$ to make I the subject.
21. $c^2=a^2+b^2$ to make b the subject.

1.9 Angles and their notation

An angle is formed when two lines meet to produce a corner; i.e. it exists in the space between two lines which are inclined to each other. If a line OA, inclined to OB, meets at O, then an angle AOB is formed.

Moving the line OA in an anti-clockwise direction increases the angle and when OA rotates to a point where it coincides with OB, then the line OA will have turned through 1 revolution.

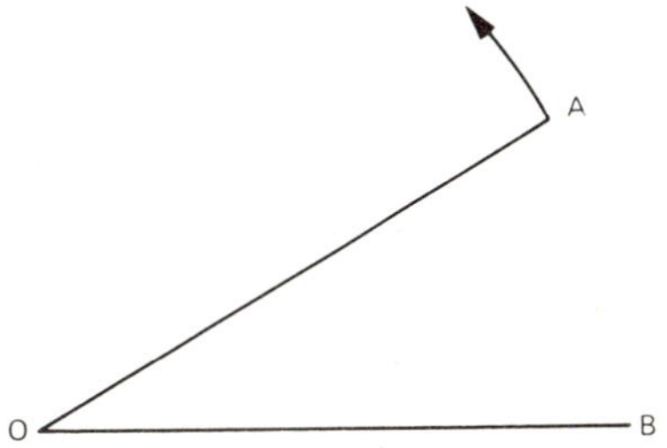

If this revolution is divided into 360 equal divisions, then each part, or $\frac{1}{360}$th of a revolution, is called a *degree* and represented by the symbol $^\circ$.

An angle of 30° is $\dfrac{30}{360}$ or $\frac{1}{12}$ of a complete revolution.

The degree unit is rather large for some purposes so it is subdivided into 60 parts; each part is called a minute and represented by the symbol $'$.

So:

60 minutes = 1 degree
360 degrees = 1 revolution

An angle can be measured by a protractor. Placing the protractor so that the corner of the angle coincides with the centre of the instrument and ensuring that OB is in line with the base of the protractor, read off the figure from the scale. In the example shown the angle is 27°.

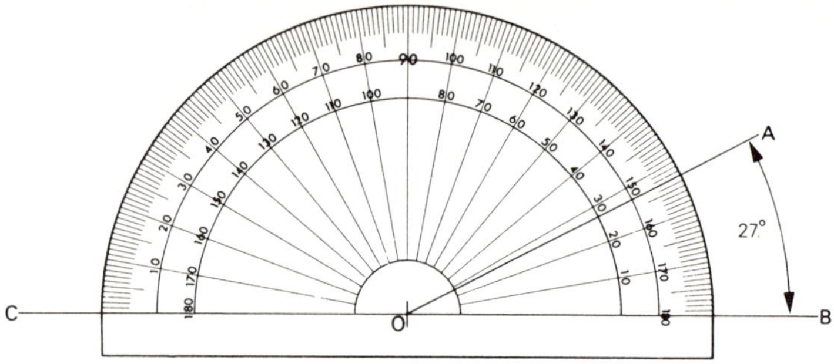

Care must be exercised to ensure that the angle stated is read from the correct scale. The scale with the zero at C must be used if angle COA is required.

Kinds of angle

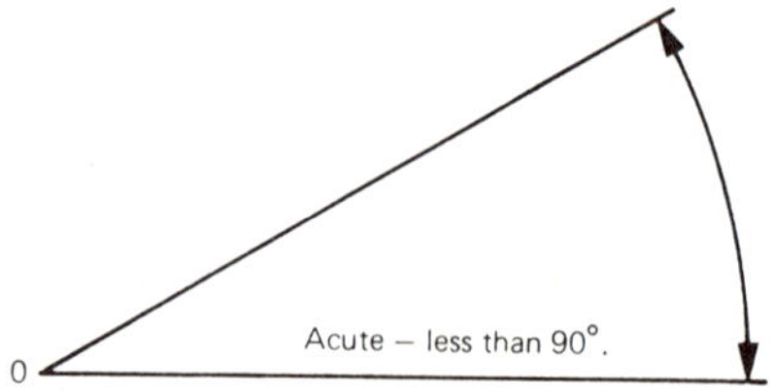

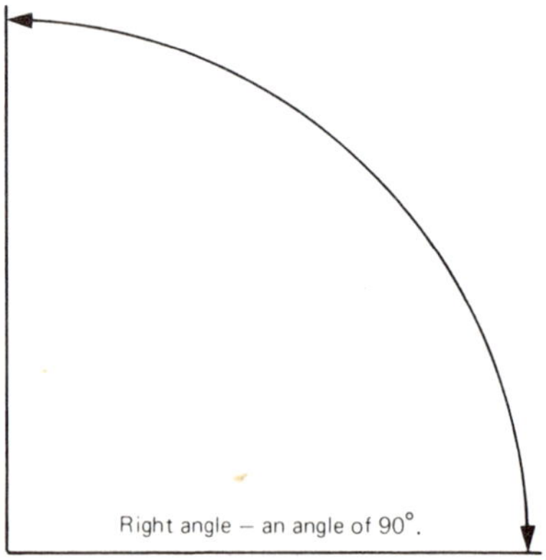

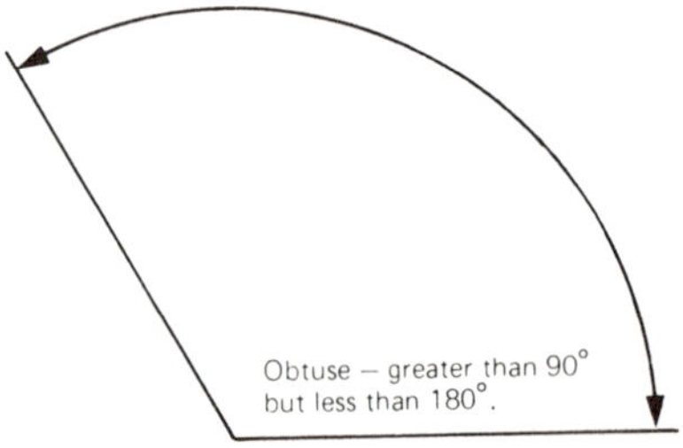

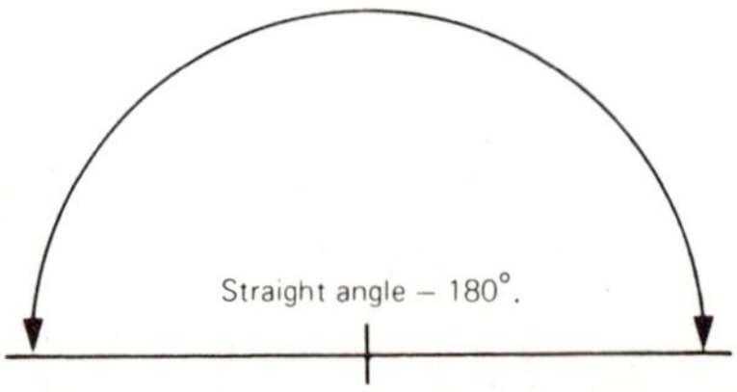

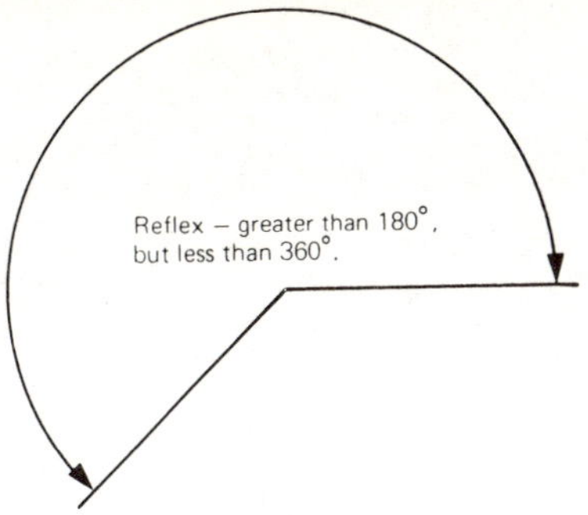

Exercises 1.9

1. A crankshaft turns through three revolutions. How many degrees does this represent?
2. The camshaft turns at half crankshaft speed. How many revolutions will the crankshaft turn when the camshaft moves through an angle of 900 degrees?
3. The contact-breaker operating cam of a coil ignition system has four lobes. If each lobe keeps the contacts open for $\frac{1}{12}$th revolution, calculate the period in degrees that the contacts are closed during one revolution of the cam.
4. A flywheel has a diameter of 420 mm. Determine the distance moved by a point on the circumference when the flywheel is turned through 9 degrees. Take π as $\frac{22}{7}$.
5. The piston of a four-stroke engine is positioned at 't.d.c., compression'. It is then moved through a reflex and straight angle. Which stroke in the cycle of operation will this new position represent?
6. Measure the acute, obtuse and reflex angles shown on page 52. State each angle in degrees.

1.10 Calculation of gear ratios

A gear ratio indicates the movement relationship between two shafts. For example, if the engine turns 5 revolutions for every 1 revolution of the propeller shaft, then the gear ratio is said to be 5:1. This indicates that the engine is turning 5 times as fast as the propeller shaft, so the movement ratio or velocity ratio is 5:1.

Moving the gear lever to engage a higher gear, e.g. 3:1, decreases the reduction so that for a given engine speed the vehicle would move faster, assuming the engine has sufficient power.

The gear ratio is governed by the relative sizes of the gear wheels; the size is generally stated as the number of teeth on the gear.

Simple gear train

When two gears of the same size mesh together the ratio is 1:1, but Fig. 1.10.1(a) shows that the direction of rotation is changed. Even if both gears were doubled in size the ratio would remain the same, the only difference being that the gear tooth loading would be reduced.

Meshing a gear of 20 teeth with a gear of 60 teeth will give a reduction when the pinion (small gear) drives the wheel (large gear). This is shown in Fig. 1.10.1(b).

In this case the pinion is the 'driver', and the wheel is 'driven'.

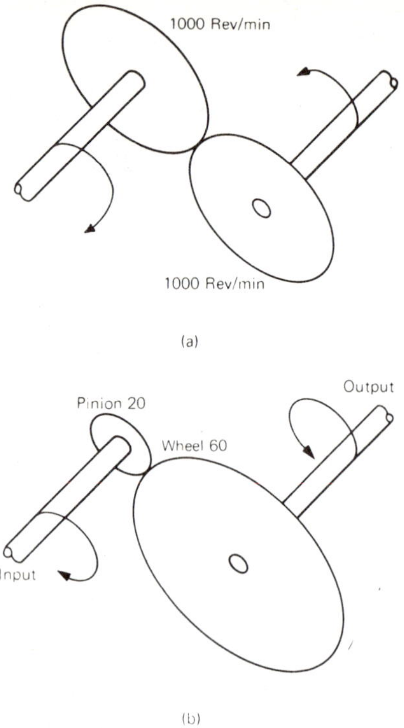

Fig. 1.10.1

1 revolution of the pinion turns the wheel through 20 teeth or $\frac{1}{3}$ revolution, so the ratio is $1:\frac{1}{3}$ or 3:1.

The ratio could also be found by using the expression

$$\text{Gear ratio} = \frac{\text{number of teeth on driven wheel}}{\text{number of teeth on driving wheel}} \text{ or } \frac{\text{driven}}{\text{driver}}$$

(Think of the poor driver after he has been driven over!)

Speed of the output shaft is $\frac{\text{input speed}}{\text{gear ratio}}$, so that speed of the output shaft would be:

$$\frac{1000}{3} = 333\tfrac{1}{3} \text{ rev/min}$$

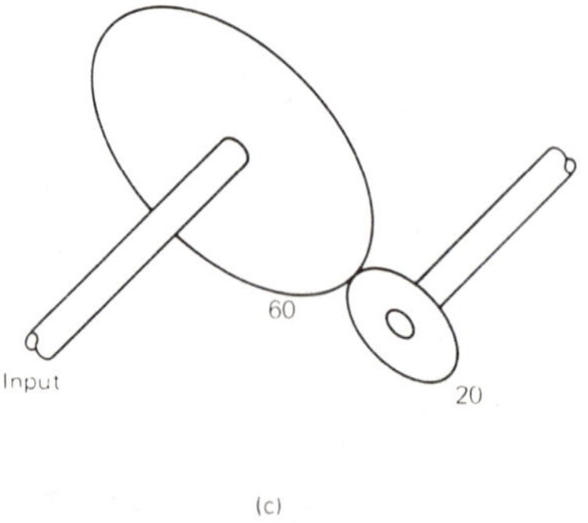

Fig. 1.10.2

Rearranging the two gears to the form shown in Fig. 1.10.2 illustrates that

$$\text{ratio} = \frac{\text{driven}}{\text{driver}} = \frac{20}{60} = \frac{1}{3} \text{ or } 1:3$$

The output speed is now three times input speed since:

$$\text{Output speed} = \frac{\text{input speed}}{\text{gear ratio}} = \frac{1000}{\frac{1}{3}} = \frac{1000 \times 3}{1}$$

$$= 3000 \text{ rev/min}$$

Compound gear train

Gearboxes are normally of the double reduction type; i.e. the gear ratio is obtained in two stages. This reduces the size of the gear wheels and restores the original direction of rotation.

Fig. 1.10.3 shows a compound train similar to that used in a gearbox. In this case the ratio of gears A and B is 2:1, and 5:1 for gears C and D.

$$\text{Speed of layshaft} = \frac{\text{input shaft speed}}{\text{ratio}} = \frac{1000}{2}$$

$$= 500 \text{ rev/min}$$

$$\text{and speed of output shaft} = \frac{\text{layshaft speed}}{\text{ratio}} = \frac{500}{5}$$

$$= 100 \text{ rev/min.}$$

The speed ratio between input and output = 1000:100
= 10:1

An easy method for obtaining the gear ratio of a compound gear train is to use the expression:

$$\text{Gear ratio} = \frac{\text{driven}}{\text{driver}} \times \frac{\text{driven}}{\text{driver}}$$

N.B. Both ratios are *multiplied* together.

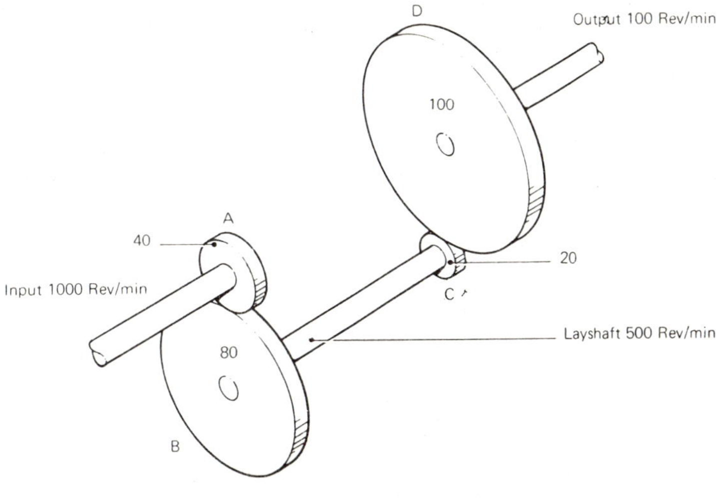

Fig. 1.10.3

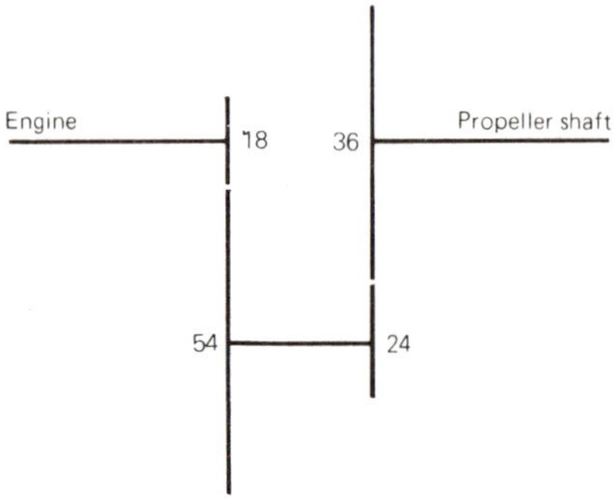

Fig. 1.10.4

Example 1

A sliding mesh type gearbox (Fig. 1.10.4) has a constant mesh pinion and wheel with 18 and 54 teeth and first-speed gears having 24 and 36 teeth, the smaller of the first gears being attached to the layshaft. Find the gearbox ratio and determine the speed of the propeller shaft when the engine is turning at 3600 rev/min.

$$\text{Gear ratio} = \frac{\text{driven}}{\text{driver}} \times \frac{\text{driven}}{\text{driver}}$$

$$= \frac{\overset{3}{\cancel{54}}}{\cancel{18}} \times \frac{\overset{3}{\cancel{36}}}{\underset{2}{\cancel{24}}}$$

$$= \frac{9}{2} = 4\tfrac{1}{2} : 1$$

$$\text{Speed of propeller shaft} = \frac{\text{input speed}}{\text{gear ratio}} = \frac{3600}{4\frac{1}{2}} = \frac{3600}{\frac{9}{2}}$$

$$= \frac{3600 \times 2}{9} = 800 \text{ rev/min}$$

Example 2

The following results was obtained during an experiment to find the ratio of a steering gearbox:

Angle turned by input shaft (steering mast) = 250°
Angle turned by output shaft (rocker shaft) = 20°

$$\text{Gear ratio} = \frac{\text{amount moved by input}}{\text{amount moved by output}} = \frac{25\cancel{0}}{2\cancel{0}} = 12\tfrac{1}{2} : 1$$

Example 3

A car has a second-speed gearbox ratio of 2·4 and a final drive ratio of 5·3. What is the speed of the road wheel when second gear is engaged and the engine is rotating at 3816 rev/min?

$$\text{Overall gear ratio} = \text{g/b ratio} \times \text{f/d ratio}$$

$$= 2{\cdot}4 \times 5{\cdot}3$$

$$= 12{\cdot}72$$

$$\text{Speed of wheel} = \frac{\text{engine speed}}{\text{gear ratio}}$$

$$= \frac{3816}{12{\cdot}72}$$

$$= 300 \text{ rev/min}$$

Further examples are shown in F. of M.V.T.

Exercises 1.10

1. A pinion of 15 teeth drives a gear wheel of 75 teeth. (a) What is the gear ratio? (b) State the speed of the output shaft if the input is turning at a speed of 2500 rev/min.

2. A pinion meshing with a wheel having 72 teeth drives the output shaft at a $\frac{1}{8}$ speed of input. How many teeth are formed on the pinion?
3. A gearbox using a compound gear train employs a driving pinion with 15 teeth and a wheel with 45 teeth. These drive a layshaft pinion with 12 teeth which meshes with a wheel having 48 teeth. Determine the gear ratio and state the output speed when the input is turning at 3600 rev/min.
4. The constant mesh gears in a gearbox have 24 and 36 teeth and the second-speed gears have 25 and 35 teeth.
   (a) Calculate the gear ratio; (b) Find the speed of the propeller shaft when the engine is turning at 4200 rev/min.
5. When the engine is turning at 3300 rev/min, the propeller shaft rotates at 600 rev/min. If the reduction between first-speed gears is 2·2:1, calculate the size of the constant mesh pinion which is required to mesh with a wheel having 60 teeth.
6. A conventional three-speed sliding mesh gearbox has a second gearbox ratio of 2·94:1. If the movement ratio between engine and layshaft is 1·4:1, what is the ratio between layshaft and mainshaft when second gear is engaged? (C & G)
7. An engine drives through a gearbox and final drive having ratios of 2·1:1 and 4·3:1 respectively. How fast will the crankshaft rotate if the rear wheels are turning at 200 rev/min? (C & G)
8. When a vehicle, fitted with a four-stroke engine, is turning a certain corner the ratio of speeds of the two driving wheels is 5:3. If the overall gear ratio is 4·5:1, calculate the angle turned by the inner driving wheel when the engine camshaft turns through 270 degrees.

1.11 Right-angled triangles and square roots

Types of triangle

A triangle is a three-sided enclosed figure. The three main types of triangle are:

Isosceles – two equal length sides.

Equilateral – three equal length sides giving three interior angles of 60°.

Right-angled – one interior angle of 90°.

The sum of the interior angles of any triangle is 180°.

Theorem of Pythagoras

Named after the Greek mathematician Pythagoras, this theorem states the relationship between the sides of a right-angled triangle.

Pythagoras showed that *the square of the hypotenuse is equal to the sum of the squares of the other two sides.*

Representing the hypotenuse (side opposite the right angle) by the symbol c and the other sides by a and b, and drawing a square on each side as shown in Fig. 1.11.1(a),

$$c^2 = a^2 + b^2$$

In Chapter 1.7. it was stated that sides of 3, 4 and 5 units form a right-angled triangle. This can be verified by the theorem

$$c^2 = a^2 + b^2$$

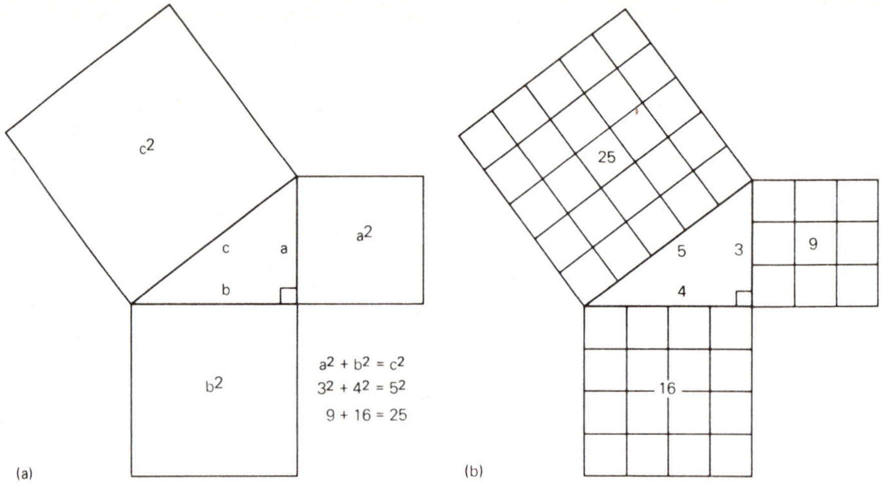

Fig. 1.11.1

Written in simple form this means

$$c \times c = a \times a + b \times b$$

If $a = 3$ and $b = 4$ then

$$c \times c = 3 \times 3 + 4 \times 4$$
$$c \times c = 9 + 16$$
$$c \times c = 25$$
$$c = 5 \text{ since } 5 \times 5 = 25$$

Drawing the squares on each side of a 3, 4, 5 triangle illustrates the theorem (Fig. 1.11.1(b).

Remember that the Theorem of Pythagoras only applies to *right-angled triangles.*

This proves the previous statement and provides a method of constructing a right-angled triangle.

Before attempting to solve further problems, it is advisable to establish a method for finding the square and square root of a number.

Square and square root

The square of a number is the number multiplied by itself. For example, the square of 4 is written as 4^2; read as '4 to the power of 2' or '4 squared', and means 4 x 4.

Working this system in reverse gives the square root. The square root of 16 is the number which, when multiplied by itself, gives 16. The square root of 16 is written as $\sqrt{16}$.

Examples

1. $2^2 = 4$
2. $3^2 = 9$
3. $4^2 = 16$
4. $5^2 = 25$
5. $6^2 = 36$
6. $7^2 = 49$
7. $\sqrt{144} = 12$
8. $\sqrt{121} = 11$
9. $\sqrt{100} = 10$
10. $\sqrt{81} = 9$
11. $\sqrt{64} = 8$
12. $\sqrt{49} = 7$

The square of a number can be found by multiplication, but determination of the square root of some numbers takes considerable time and effort. Tables have been constructed to simplify the task.

Examples on Square Roots

Using an extract from the square root tables (shown below), the following square roots can be ascertained.

	0	1	2	3	4	5	6	7	8	9	Mean difference 123	456	789
3·3	1·817	1·819	1·822	1·825	1·828	1·830	1·833	1·836	1·838	1·841	011	112	222
3·4	1·844	1·847	1·849	1·852	1·855	1·857	1·860	1·863	1·865	1·868	011	112	222
3·5	1·871	1·873	1·876	1·879	1·881	1·884	1·887	1·889	1·892	1·895	011	112	222
3·6	1·897	1·900	1·903	1·905	1·908	1·910	1·913	1·916	1·918	1·921	011	112	222
3·7	1·924	1·926	1·929	1·931	1·934	1·936	1·939	1·942	1·944	1·947	011	112	222
3·8	1·949	1·952	1·954	1·957	1·960	1·962	1·965	1·967	1·970	1·972	011	112	222
3·9	1·975	1·977	1·980	1·982	1·985	1·987	1·990	1·992	1·995	1·997	011	112	222

Example 13

Find the value of $\sqrt{3{\cdot}6}$.

Method – look down the first column for 3·6 and read off the root under the '0' vertical column.

Answer 1·897.

Example 14

Find the value of $\sqrt{3{\cdot}67}$.

Method – the first column locates the 3·6 and the column across the top enables the 7 to be taken into account. The answer is found at the intersection between the 3·6 line and the 7 columns.

Answer 1·916.

Example 15

Find the value of $\sqrt{3{\cdot}678}$.

Method – proceed as in the previous example to find the root of 3·67. The effect of the figure in the third decimal place is shown under the 'difference columns'. At the intersection between the horizontal '3·6' column and the vertical '8' difference column is shown the value 2. This must be added to the previous value 1·916.

$$\therefore \sqrt{3{\cdot}678} = 1{\cdot}916 + 0{\cdot}002 = 1{\cdot}918$$

Position of decimal point

Any number in the range 1 to 100 for which the square root is required has the decimal point clearly shown in the table. For numbers outside these limits a special method is used where it is difficult to establish by inspection the point position.

Example 16

Find the value of $\sqrt{372{\cdot}6}$.

The first step is to separate the digits in groups of two to the left and right of the decimal point, i.e.

$$3\ 72{\cdot}60$$

To find the square root the table is consulted – the first group indicates the section of the table to use – in this case 3. The significant figures obtained from the table for the number 3·726 is

$$1931$$

Writing this down under the original value:

$$\begin{array}{cccc} 3 & 72 & \vdots & 60 \\ \downarrow & \downarrow & \downarrow & \downarrow \\ 1 & 9 & \cdot & 3\ 1 \end{array} \qquad \therefore \sqrt{372{\cdot}6} = 19{\cdot}31$$

Example 17

Find the value of $\sqrt{3726}$.

$$\begin{array}{ccc} 37 & 26 & \vdots \\ \downarrow & \downarrow & \downarrow \\ 6 & 1 & \cdot\ 0\ 4 \end{array} \qquad \therefore \sqrt{3726} = 61{\cdot}04$$

Example 18

Find the value of $\sqrt{0{\cdot}000\ 342}$.

$$\begin{array}{cccccc} 0 & \vdots & 00 & 03 & 42 & \\ \downarrow & \downarrow & \downarrow & \downarrow & \downarrow & \\ 0 & \cdot & 0 & 1 & 8 & 4\ 9 \end{array} \qquad \therefore \sqrt{0{\cdot}000\ 342} = 0{\cdot}018\ 49$$

Examples on Theorem of Pythagoras

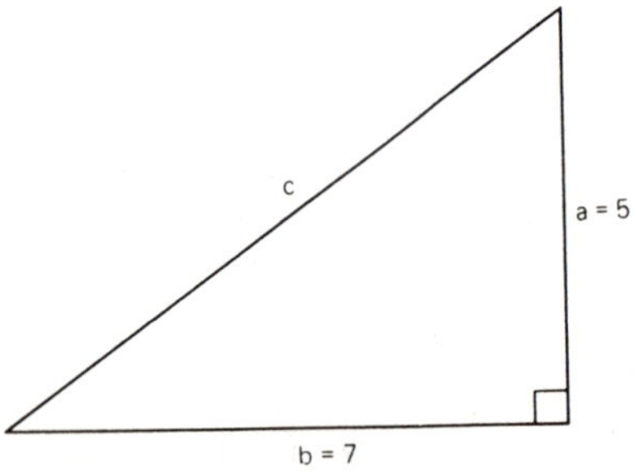

Example 19

Find the length of the hypotenuse if the length of the other sides are 5 cm and 7 cm.

Theorem of Pythagoras states:

$$c^2 = a^2 + b^2$$

so

$$\begin{aligned} c^2 &= 5^2 + 7^2 \\ c^2 &= 25 + 49 \\ c^2 &= 74 \\ c &= \sqrt{74} \\ c &= 8{\cdot}602\,\text{cm} \end{aligned}$$

Example 20

The shortest and longest sides of a right-angled triangle are 4·5 m and 6·7 m respectively. Find the length of the other side.

$$c^2 = a^2 + b^2$$
$$6{\cdot}7^2 = 4{\cdot}5^2 + b^2$$
$$44{\cdot}89 = 20{\cdot}25 + b^2$$
$$44{\cdot}89 - 20{\cdot}25 = b^2$$
so $$b^2 = 24{\cdot}64$$
$$b = \sqrt{24{\cdot}64}$$
$$b = 4{\cdot}962 \text{ m}$$

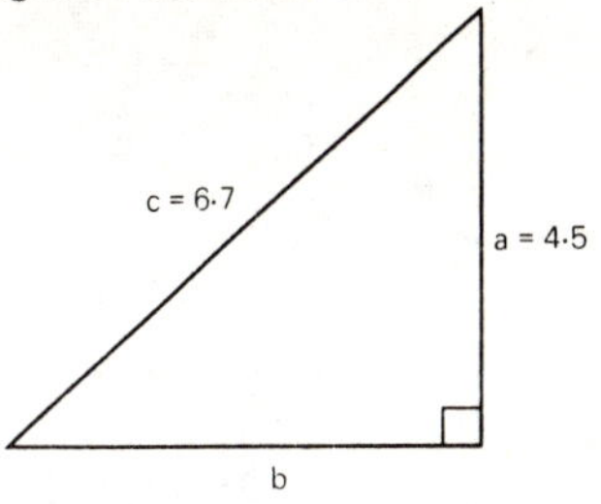

Exercises 1.11

Find the square root of:

1. 19. 2. 27·2. 3. 6·243. 4. 98·67

Find the square of:

5. 5·2. 6. 7·24. 7. 8·125. 8. 0·52

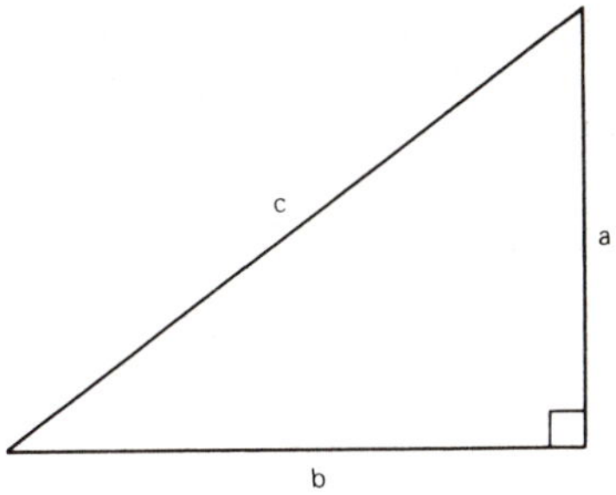

Questions 9—12 refer to a right-angled triangle. Find the third side.

9. a = 5; b = 6; c = ?
10. a = 3·23; b = 2·1; c = ?
11. a = 1·3; b = ? c = 6·8
12. a = ? b = 3·42; c = 9·54
13. An engine has a stroke of 50 mm and a connecting rod 100 mm long. Calculate the distance between the main bearing and gudgeon pin centres when the crank is 90° past the t.d.c. position.
14. How far has the piston moved down the cylinder in question 13?
15. An engine has a stroke of 100 mm and a connecting rod 200 mm long. Calculate the distance between the main bearing and gudgeon pin centres when the crank and connecting rod centre lines form a right angle.
16. A rectangular sub-frame of a chassis has an overall length and width of 1·9 m and 0·9 m respectively. Calculate the length of a diagonal line drawn between extreme corners.
17. A propeller shaft has a length of 2·4 m measured between the centres of the universal joints. On loading the vehicle the rear joint moves vertically through a distance of 0·2 m so that when the vehicle is fully laden the propeller shaft is horizontal. Find the amount that the propeller shaft has to reduce in length during the loading period.

1.12 Simple graphs

A graph is a diagram which shows the relationship between two quantities. It enables an observer to see how the variation in one quantity affects the value of the other quantity.

Consider typical results obtained from an experiment to show the relationship between Celsius and Fahrenheit temperature scales.

Most probably these would be shown in tabulated form:

C	10	15	20	25	30
F	50	59	68	77	86

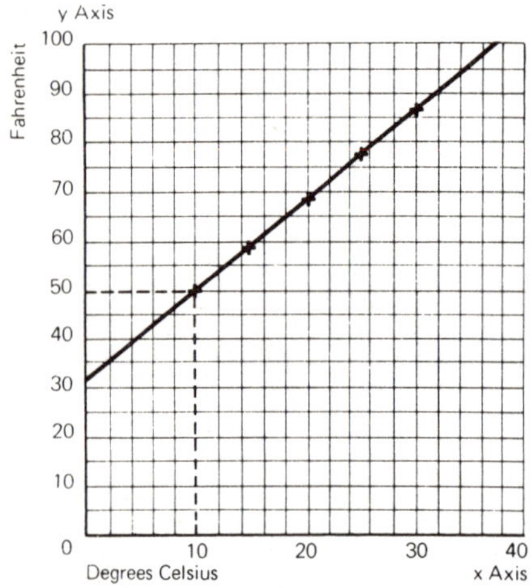

Plotting these results in graphical form:

This graph is constructed by using the following method.

1. On squared paper a horizontal line, called the x axis, is drawn. This is divided up into equal portions to represent the C temperature and the scale is stated by marking in the various numbers, for example 0, 10, 20, 30. The scale selected should be simple to read and awkward units such as 3 and 7 should be avoided.
2. At point O, called the origin of the graph, a vertical line or y axis is drawn. This will represent the Fahrenheit scale. Mark scale on axis as before and state the name of the quantity on each axis, that is °F and °C.
3. The first set of readings in the table show that C = 10 when F = 50. To plot this, draw a vertical line through 10°C and a horizontal line through 50°F. The intersection of these two lines shows the F reading when the C reading is 10. When this plotting system is understood, the intersection point only need be shown.
4. Using the previous rule plot the remaining results.
5. Draw the graph through the various plotting points.

Having constructed the graph it will be seen that the graph can be used to obtain values which were not included in the tabulation of results, for example:

95°F = 35°C
41°F = 5°C
32°F = 0°C

Greater accuracy could be achieved if a larger scale were used. The graph should always be as large as possible. The following examples show that graphs are not always straight lines.

Example 1

In an experiment to compare the discharge of water through an orifice, with the pressure head, the following results were achieved.

Pressure head (mm)	100	200	300	400	500	600	700	800
Discharge (litres/hour)	6·5	11	13	14	17	17·5	19·5	20

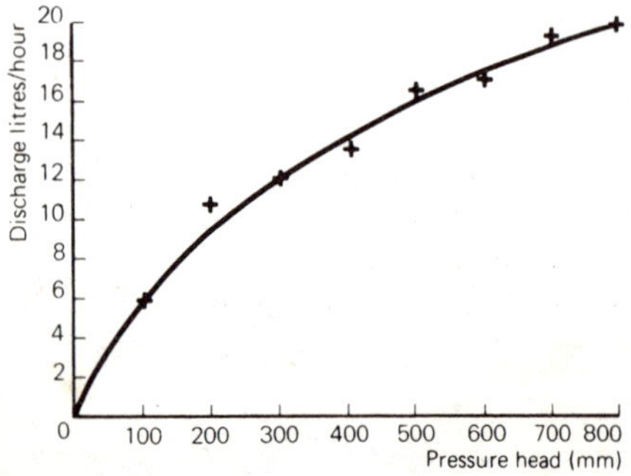

Plot these results.

Plotting these values show that slight errors in the results occur since the points do not fall in a definite pattern. To overcome this, the curve is drawn through the mean of the points.

Example 2

The lift or displacement of an engine valve in relation to crankshaft movement is shown by the following table. From the figures obtained plot the graph of valve lift against crankshaft movement.

Crankshaft movement (Degrees)	0	20	40	60	80	100	120	140	160	180	200	220	240
Valve lift (mm)	0	1	3	7	11	13	14	13	11	7	3	1	0

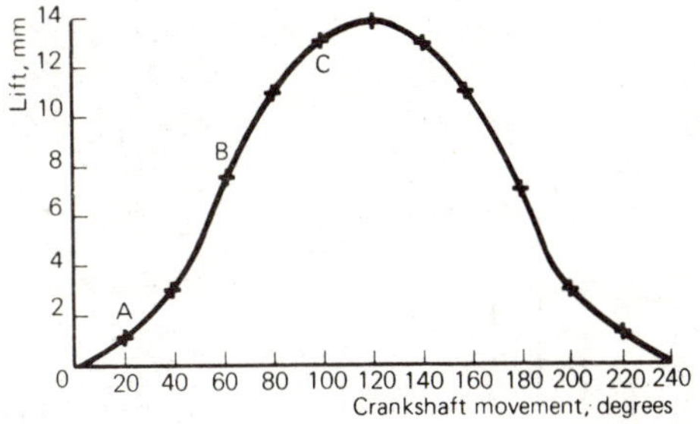

The first quantity stated in the table is generally plotted on the Ox axis. In a similar way the subject following the word 'against', or any subject based on 'time' is also plotted on the Ox axis.

The graph above shows how the valve is gradually raised from its seat at A, rapidly opens at B, and slows down at C as it nears the fully open position. A similar motion occurs during the closing period. The graph may be used to show the effect of tappet clearance on the valve period.

Example 3

During a vehicle performance test the following results were obtained.

Speed (km/h)	0	24	56	72	88	96	112	116	120
Time (Seconds)	0	1	2	3	4	5	6	7	8

Plot a graph of speed and time and use this graph to determine the speed at $2\frac{1}{2}$ seconds after starting.

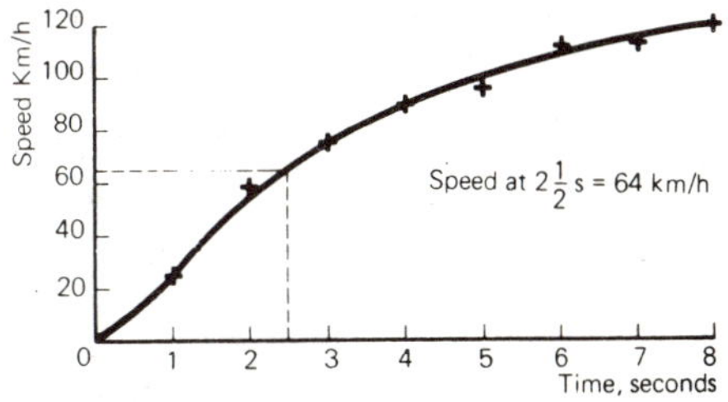

The previous examples show a definite relationship between the two quantities. When unrelated quantities are plotted, a *chart* is formed — this is often used for statistical purposes. Often these unrelated facts can only be shown by a series of vertical lines; the variation in these values often make it impossible to draw a definite line or curve. The temperature chart of the foot of a hospital bed is a good example of this form of graph.

Example 4

The monthly sale by the Parts Department of radiator caps for the period Jan-Dec was as follows:

12, 3, 8, 11, 19, 25, 27, 24, 12, 7, 18 and 2.

Draw a chart showing the monthly sales.

This type of question may be answered by drawing one of the following forms of chart.

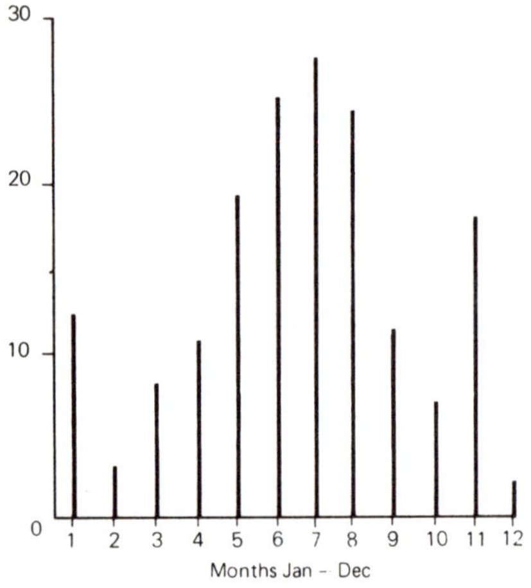

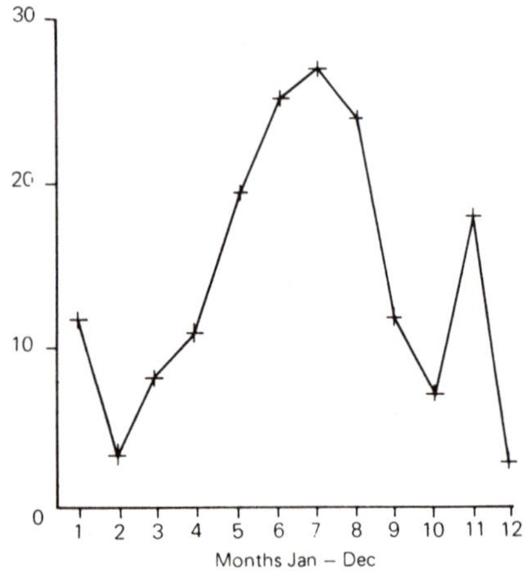

Exercises 1.12

1. The following results were obtained during an experiment on friction.

Mass kg	1	2	3	4	5
Frictional force newtons	3·1	7·2	10·4	13·2	16·8

Plot these values and from your graph find the frictional force when the mass is 2·75 kg.

2. The table shows the current flowing in a circuit for a given voltage.

Voltage	2	4	6	8	10	12
Current A	1·3	2·7	4	5·3	6·7	8

Plot a graph of current against voltage and determine the voltage required to produce a current flow of 6 A.

3. Power output (kilowatts) of an engine in relation to speed is shown by the table below.

Speed (rev/min)	1000	2000	3000	4000	5000
Brake power kW	6	14·2	21	25·4	27

Plot these results and state the expected power output at 2500 rev/min.

4. Using square root tables plot a graph to show the square root of numbers up to 50. Use this graph to find $\sqrt{20}$.

5. The effect of pressure on the boiling point of water is given by the following table:

Pressure (kN/m^2)	76	103	138	172	207	241
Boiling point (°C)	90	100	108	115	121	126

What would be the boiling point of the water in a cooling system if it was pressurised to a value of 148 kN/m^2?

1.13 Drilling speeds

To ensure maximum tool life as well as taking the shortest time to do a job, the drilling speed and feed recommended for the tool should be used. Generally spindle speeds on bench drilling machines can be varied to suit the size and type of drill and the material of the job, whereas the speed of a portable hand drill has to satisfy the range of drills which fit the chuck.

Calculation of spindle speeds

In many ways the action of a drill is similar to that of a cutting tool in a lathe.

Consider a bar of diameter d mm mounted in a lathe which has a spindle speed of n rev/min.

If S is the maximum linear speed (surface speed) of the bar, then:

$$S = \text{circumference} \times \text{rotation speed}$$
$$S = \pi d \times n \text{ mm/min}$$
$$S = \frac{\pi d n}{1000} \text{ m/min}$$

It will be seen that S is the speed at which the bar passes the tool, so this symbol also represents the cutting speed.

Fig. 1.13.1 shows the end elevation of a twist drill; the two cutting edges will be recognised. Consideration of the linear speed of one cutting edge reveals that the speed is proportional to the radius. The length of the arrow in Fig. 1.13.1 represents the speed, and this shows that the maximum cutting speed occurs at C. The formula used for a lathe is also suitable for a drill.

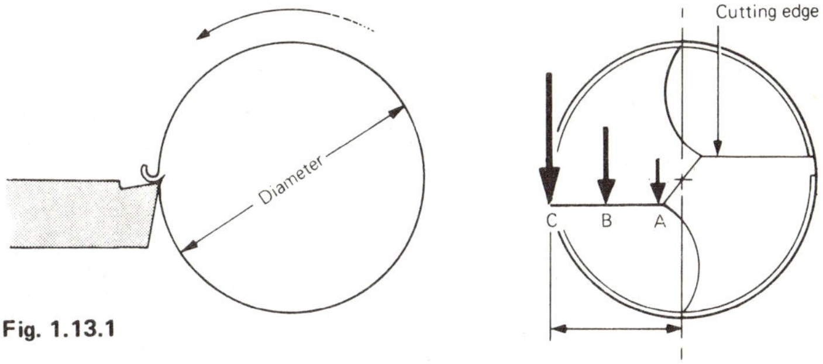

Fig. 1.13.1

Example 1

For drilling mild steel a cutting speed of 30 m/min is recommended. At what speed should a 6 mm diameter drill be rotated?

$$S = \frac{\pi d n}{1000}$$
$$1000\,S = \pi d n$$
$$n = \frac{1000\,S}{\pi d}$$
$$n = \frac{1000 \times \overset{5}{\cancel{30}}}{\frac{22}{7} \times \cancel{6}} = \frac{\overset{2500}{\cancel{5000}} \times 7}{\underset{11}{\cancel{22}}} = \frac{17\,500}{11}$$

$$n = 1591 \triangleq 1600 \text{ rev/min}$$

Example 2

A 12 mm diameter drill rotates at 1000 rev/min. Is this a suitable speed for cutting mild steel?

$$S = \frac{\pi dn}{1000} = \frac{22 \times 12 \times 1000}{7 \times 1000} = \frac{264}{7} \triangleq 38 \text{ m/min}$$

Since 30 m/min is the recommended cutting speed, the spindle speed should be lowered to 800 rev/min.

Average cutting speed for high-speed drills.

	Brass	Soft steel	Cast Iron	Bronze
m/min	60	30	24	23

Drilling feed

Drilling feed is the distance the drill advances into the work per revolution. This is normally controlled by the operator, either by moving the feed handle or, as in the case of portable equipment, by direct effort applied to the drilling machine.

A high-speed drill rotating at the recommended cutting speed may be given a feed of

(*0·02 mm per revolution per 1 mm of drill diameter*)

Therefore, 2mm diameter drill requires a feed of 0·04 mm/revolution, 5 mm diameter drill requires a feed of 0·1 mm/revolution.

Exercises 1.13

Determine the spindle speed for the following:

1. Drilling cast iron with a 6 mm diameter drill.
2. Drilling brass with a 12 mm diameter drill.
3. Drilling mild steel with a 2 mm diameter drill.
4. Drilling mild steel with a 5 mm diameter drill.
5. Drilling bronze with a 7 mm diameter drill.
6. What feed and speed should be given to a 12-mm diameter drill which is cutting a hole in a conventional cylinder block?
7. Calculate the cutting speed for mild steel in metres/second.
8. A drill of diameter 20 mm rotating at the correct speed is given a feed of 0·38 mm/revolution. Is this feed suitable for this drill?

Cutting action of tools

Most of us at some time have attempted to cut or drill metal with a blunt tool. Often this results in poor quality work and a considerable loss of energy — both physical and mental.

If the reader understands the basic principle underlying the way in which a tool cuts a material, then this can be applied to a wide range of tools.

Fig. 1.13.2 shows the cutting edge of a tool which could be a chisel, drill, hacksaw blade or any other cutting tool. The cutting action is obtained by forcing a sharp wedge into the metal and causing the top layer to break or chip from the main work piece. An

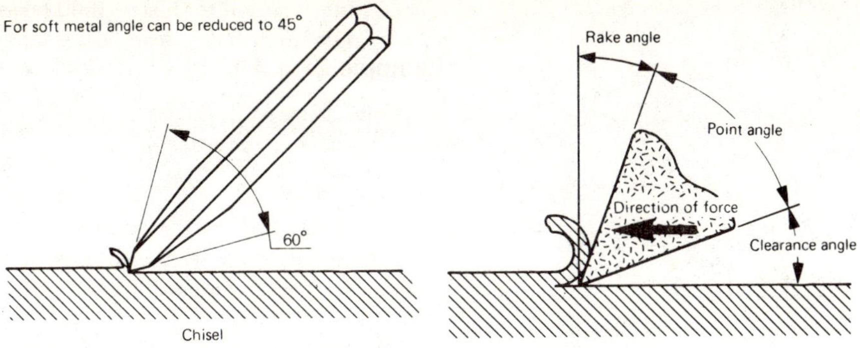

(a)

Fig. 1.13.2

efficient tool; one which requires the minimum force to give a satisfactory cut; is obtained when the tool is:

(a) made of a suitable material
(b) heat treated to give a tough and often hard cutting edge.
(c) sharpened to suit the metal being cut.

Cutting angles

The two main angles are:

Clearance angle – keeps the cutting edge in contact with the work and ensures that the force applied to the tool is concentrated at the cutting edge.

Rake angle – controls the angle of the point. A small point angle allows the tool to cut into the material easily but tools having this pointed shape are weak and soon become damaged, e.g. a wood chisel would not last long if it was used on steel.

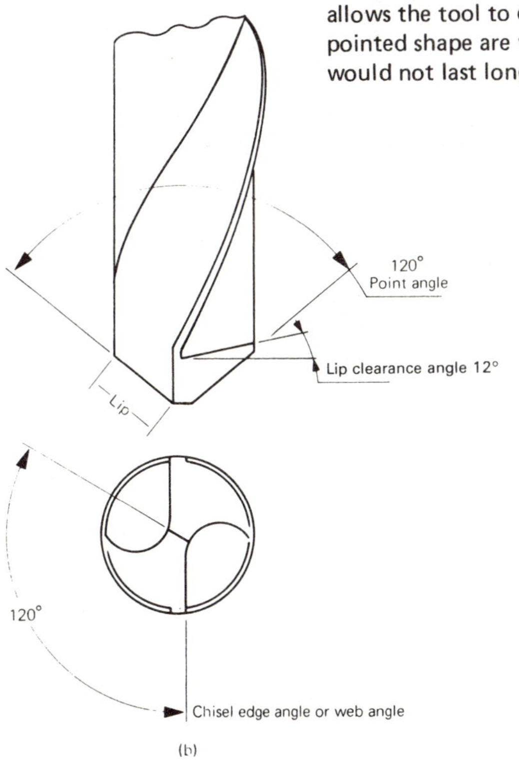

(b)

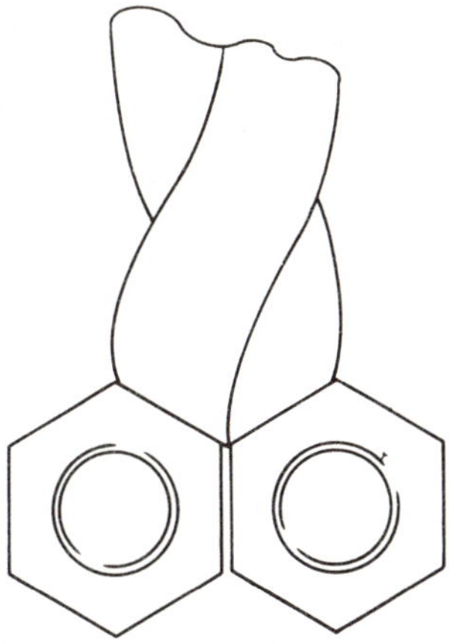

Checking point angle with two nuts

(c)

Fig. 1.13.3

Tool Angles for General Work

Cutting angles of tools such as hacksaw blades and files are set by the manufacturer but in many cases the angle is set when the tool is sharpened. Fig. 1.13.3 shows suitable angles for a drill. Besides having the correct angles on a drill, the lips must be the same length – unequal lengths will cause the drill to cut an oversize hole and vibrate in use.

1.14 Measurements—methods

Various measuring instruments are used by a motor mechanic, each instrument serving a particular purpose. In this chapter some of the basic equipment is considered, but before proceeding, a word of warning. Instruction on measuring instruments is normally aided by the student actually handling the items and it must be appreciated that if they are treated carelessly this can often ruin them. Replacement or repair is expensive so do not drop the instrument. After use remove the moisture and lightly oil the equipment to protect it from corrosion.

Engineer's rule

A rule is used to measure any flat surface, for example the length of this page, but is unsuitable for direct reading of an object such as a steel ball. The straight edges of the rule are also utilised for alignment purposes or for checking a flange for distortion.

Calipers – internal and external

Fig. 1.14.1 shows two dimensions which can be conveniently taken by calipers; the external caliper is used for the crankpin and the internal caliper for the brake cylinder. The sketch shows the manner in which they are used. Holding one leg on the crankpin, the other leg of the external caliper is slid over the pin to locate the largest dimension. A similar method can be used for the internal caliper; the setting may be checked by rocking the instrument in the direction of the arrow. Whenever the caliper has to be set to a given distance, the final adjustment is obtained by tapping one leg (Fig. 1.14.2).

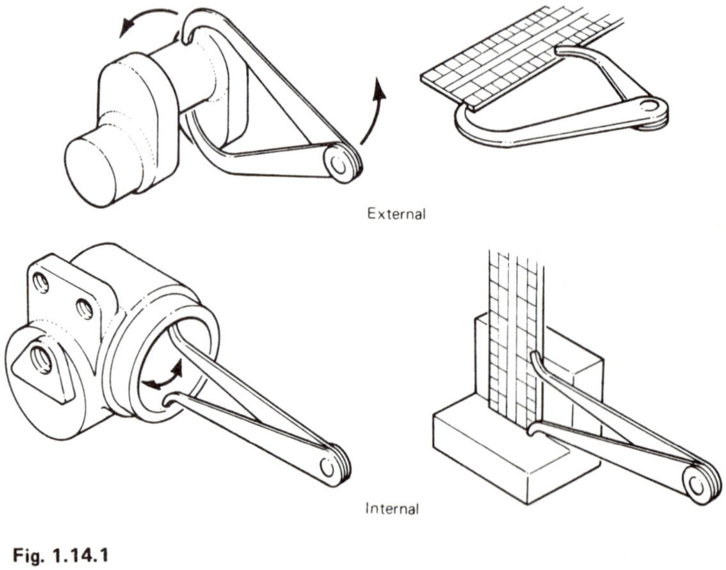

Fig. 1.14.1

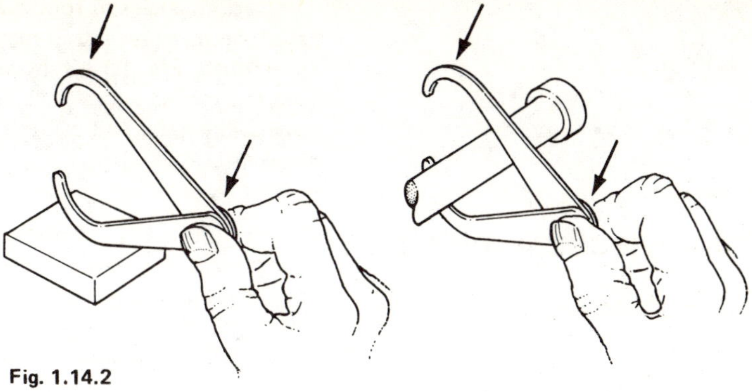

Fig. 1.14.2

Calipers are not used only for round surfaces. The thickness of a key may be checked with an external caliper, while the width of the keyway is measured with an internal caliper.

Feeler gauge

Consisting of several strips of thin steel, each marked with its thickness, this gauge is used to measure the clearance between two components. Although measurement of valve clearance is the most common application (Fig. 1.14.3), it is also used to check the end float of camshafts, crankshafts, and so on. Whenever feeler gauges are used to check the fit of circular items such as pistons, the width of the gauge should be stated.

The precise clearance can only be obtained after experience has been gained as to the effort required to slide out the blade. For this reason gauges used for piston clearance sometimes incorporate a spring balance.

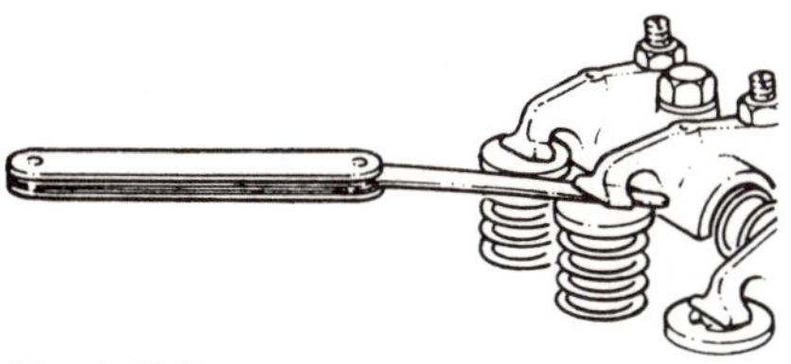

Fig. 1.14.3

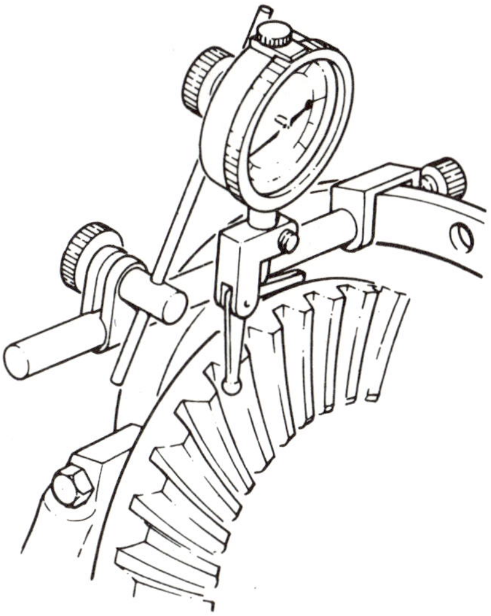

Fig. 1.14.4

Clock gauge

This dial test indicator, often abbreviated to d.t.i., is used to measure small linear movements. It consists of a clock gauge, finely calibrated in Metric units of length and a number of mounting attachments (Fig. 1.14.4).

To mount the gauge to a component, assemble the required attachments and leave all attachment screws loose until the gauge is correctly positioned. Remember to preload the gauge plunger away from its outer stop, and ensure that the movement to be measured is not greater than the stroke of the gauge plunger.

The gauge will stick if the plunger is oiled.

Cylinder gauge

Similar to the d.t.i., this instrument consists of a clock gauge and a linkage incorporated in the handle. Various distance pieces are provided to give a large range of measurement. It is sometimes called a comparator because it can only compare one distance with another.

Fig. 1.14.5 shows the gauge being used to determine the size of a cylinder. Rocking the gauge in the direction shown enables the smallest reading to be found and rotation of the bezel ring allows the gauge to be set to zero. After the number of revolutions of the needle (sometimes registered on a separate small dial) has been noted the instrument is placed in an external micrometer and the actual distance determined.

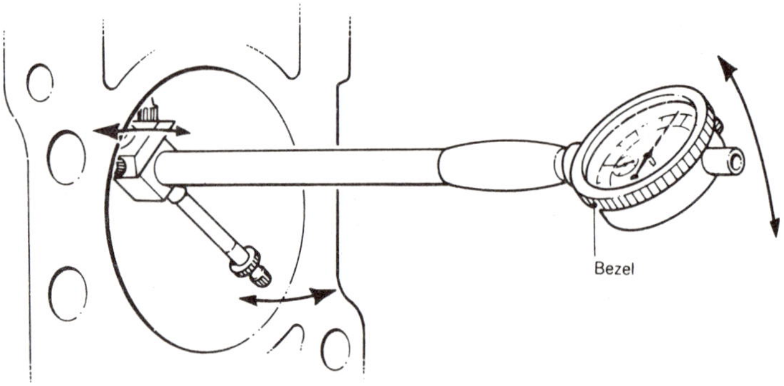

Fig. 1.14.5

Micrometer

This is a development of the caliper which enables linear measurements to be determined to an accuracy of at least 0·01 mm.

The principle can be shown by the diagrammatic sketch (Fig. 1.14.6). This represents a screw which engages with an internal thread of pitch 0·5 mm in the frame. Rotating the screw one revolution causes the screw to advance 0·5 mm. If the head of the screw is divided into fifty parts then each division will represent 0·01 mm. Therefore turning the screw to the extent of $3\frac{7}{50}$ turns will move the screw to the extent of:

$$\begin{aligned} 3 \times 0{\cdot}5 &= 1{\cdot}5 \text{ mm} \\ 7 \times 0{\cdot}01 &= \underline{0{\cdot}07} \text{ mm} \\ \text{Total} &= 1{\cdot}57 \text{ mm} \end{aligned}$$

Fig. 1.14.7 shows the layout of a typical micrometer. The frame carries an anvil and opposite this is an internal thread which engages

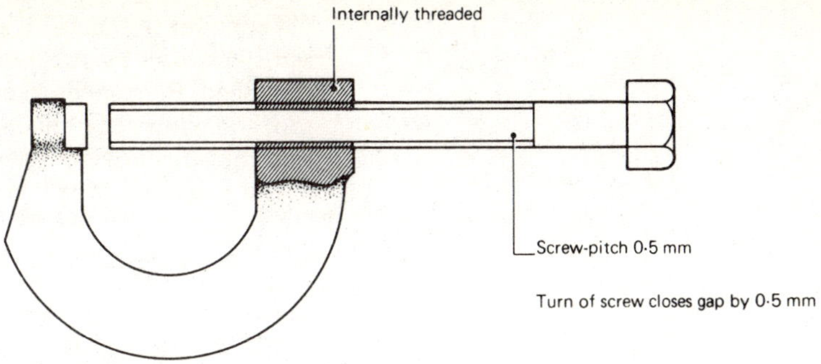

Fig. 1.14.6

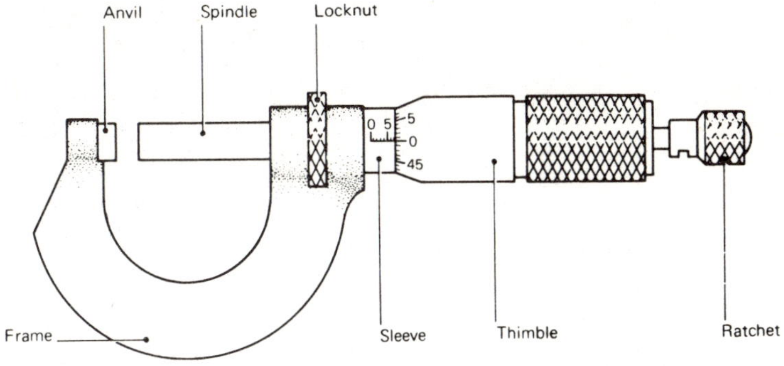

Fig. 1.14.7

the spindle. A thimble, locking up on a taper, allows the spindle to be rotated. Also attached to the frame is a scaled sleeve which is divided into 0·5 mm intervals: vertical scale markings above the horizontal datum line are spaced at 1 mm intervals whereas vertical markings below the datum indicate the 0·5 mm positions. The following examples show the manner in which the scale is read.

(a)

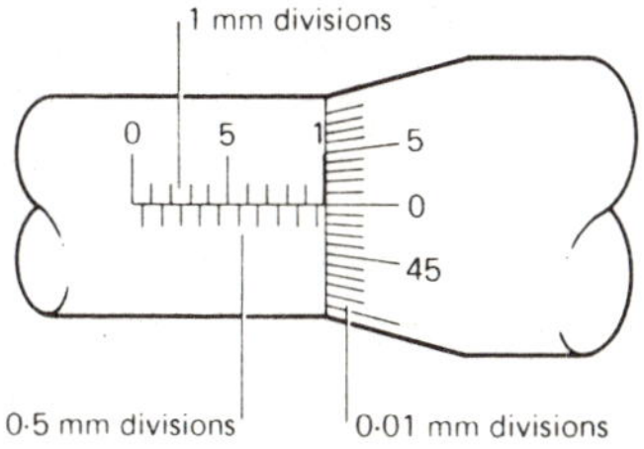

(b)

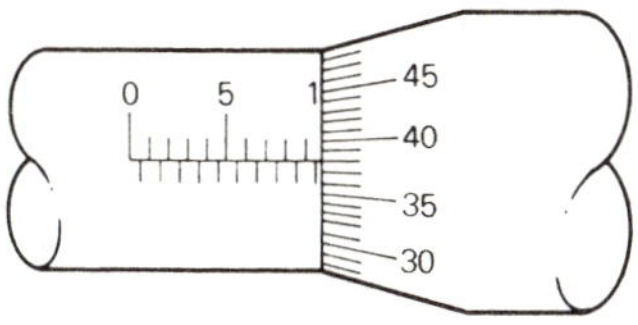

(c)

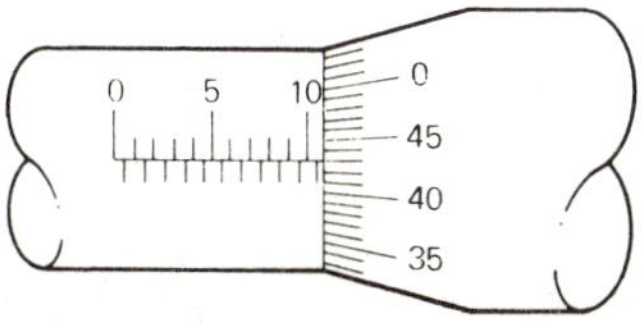

(a) 1 mm divisions

(b) 0·01 mm divisions

(c) 0·5 mm divisions

Reading (a)

Number of mm = 10 = 10·00
Number of 0·01 mm = 0 = 0·00
10·00 mm

Reading (b)

Number of mm = $9\frac{1}{2}$ = 9·50
Number of 0·01 mm = 38 = 0·38
9·88 mm

Reading (c)

Number of mm = $10\frac{1}{2}$ = 10·50
Number of 0·01 mm = 43 = 0·43
10·93 mm

Care must be exercised not to overtighten the thimble. Since the actual reading depends on the tightness, a ratchet is provided.

Normal type micrometers have a scale range of 25 mm so a number of micrometers are needed to cover the measurements of motor vehicle components. If, for example, a crankpin of 54 mm diameter is to be checked, then a 50–75 mm micrometer is needed. There are, however, some micrometers which can be fitted with a distance piece to enlarge the range.

If a micrometer is replaced in its box when not in use, it seldom requires adjustment, but the accuracy should be checked periodically. In the case of a 0–25 mm micrometer this is accomplished by checking that the reading is zero when the spindle contacts the anvil. Larger instruments are checked against a test piece of standard length or diameter.

Three adjustments are normally provided:

(a) thread adjustment – to compensate for thread wear a split nut is provided.

(b) thimble adjustment – the thimble is pulled up on a taper by a screw, so the position of the thimble scale may be set to the correct position.

(c) sleeve adjustment – rotating the sleeve with a 'C' spanner allows minute adjustment of the scale.

Internal micrometer

This micrometer measures the distance between two internal surfaces. Fig. 1.14.8 shows the instrument being used to determine the bore of an engine cylinder.

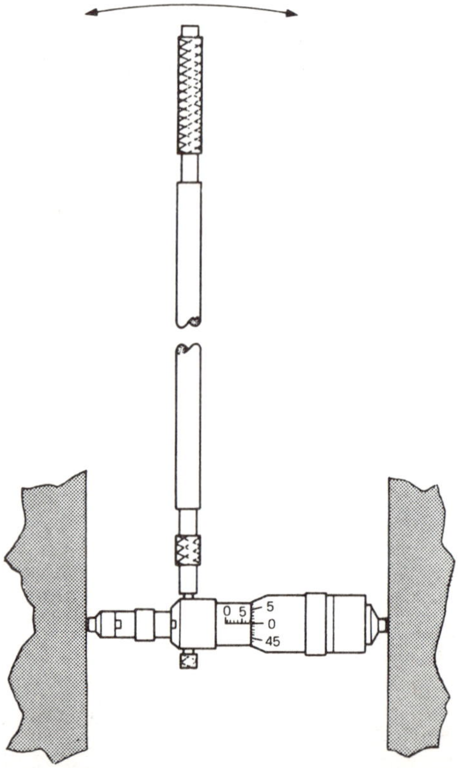

Fig. 1.14.8

The most common type has a length of 50 mm and a scale range of 10 mm. The smallest hole which it can measure is 50 mm and distance pieces of length 10, 20, 40, 60 mm, and so on, enable a wide range of sizes to be covered.

Whenever a micrometer is used, it is important to ensure that the faces are clean. Small grit traps, cut in the distance pieces of an internal micrometer, collect the dirt when the distance piece is rotated before clamping.

When this micrometer is being used, it must be rocked in the direction shown in the diagram until the correct feel is achieved. Remember that you have great leverage on the instrument and that it will be damaged if it is forced.

Since the scale is limited to 10 mm the length of distance piece must be added to the scale reading to obtain the actual dimension.

Exercises 1.14

Q1

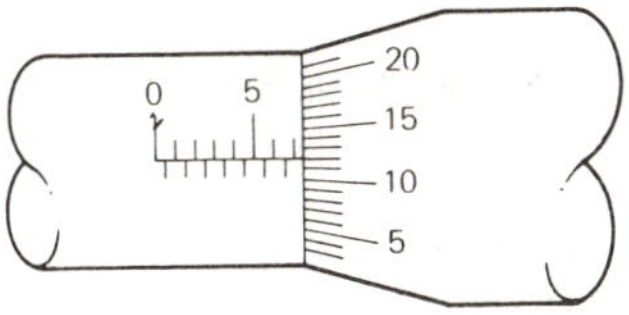

Q2

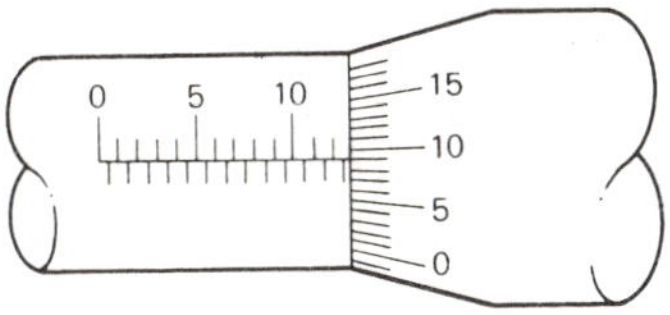

Q3

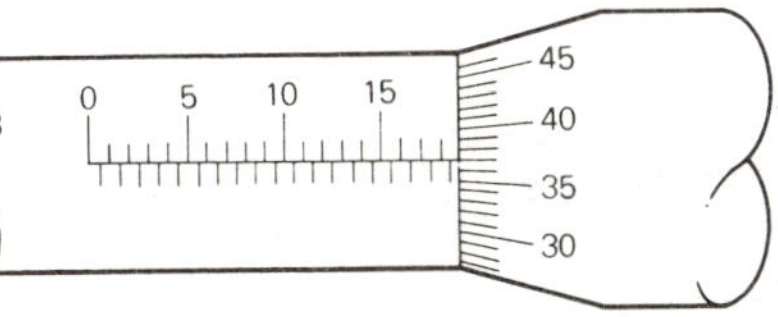

Questions 1–3 refer to a metric micrometer. State the reading in each case.

4. What is the pitch of a metric micrometer thread?
5. Draw the scale of a micrometer set to a reading of 5·64 mm.

2 SCIENCE (a) Heat

2.1 The nature of heat

A substance is made up of tiny particles called molecules and these consist of even smaller particles called atoms. Observations of these tiny particles shows that each one is vibrating; the amount of vibration depending on the 'hotness' of the substance. So,

heat is internal energy and is capable of doing work.

Heat engines obtain their power by burning fuel – the chemical energy in the fuel is 'released by ignition to make the gas very hot. This causes a force which pushes the piston down the cylinder.

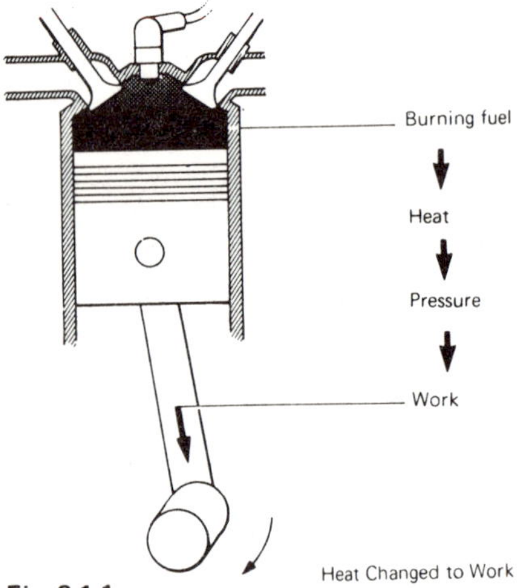

Heat Changed to Work

Fig. 2.1.1

Heat flows from a hot to a cold surface

A hot object placed beside a cold surface will cause the heat to 'flow' until the two things are at the same heat.

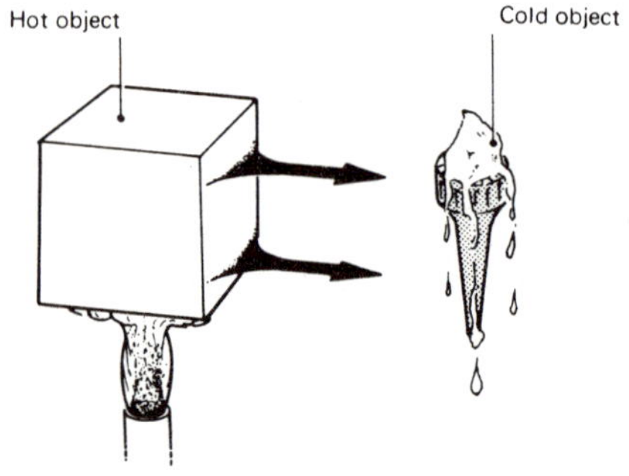

Heat Flows From a Hot Object to a Cold Object

Fig. 2.1.2

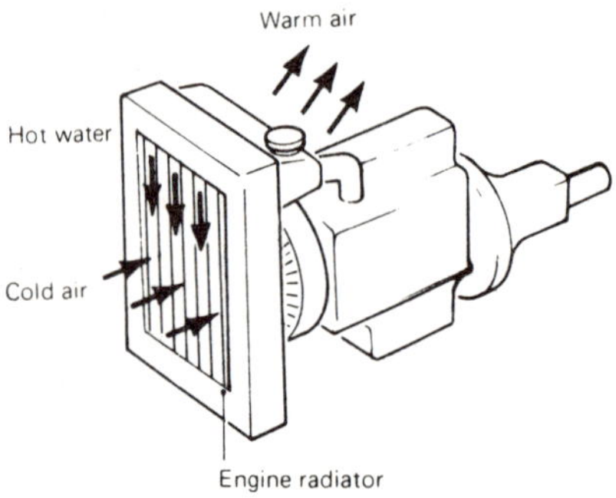

Hot Water Gives Up Heat to Cold Air

Fig. 2.1.3

2.2 Temperature

The temperature of a body is

the degree of hotness

This is different to the heat contained in an object, e.g. more energy will be required to heat four litres of water than that used to heat one litre to a similar temperature.

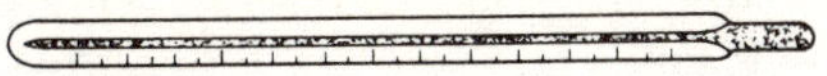

Figure 2.2.1

Although terms like 'hot' and 'cold' give an indication of the temperature, they lack accuracy and can lead to misunderstanding, e.g. A hot day means something different to a hot engine.

Temperature is measured by a thermometer

Most substances grow bigger (or expand) as they are heated and this feature is used in the common type of thermometer. This type consists of a thin, sealed glass tube having a bulb containing mercury or alcohol at one end.(Fig. 2.2.1). When the mercury is heated it expands and flows along the fine bore. The temperature is read by noting the position, relative to a scale, of the end of the thin column of the liquid. Mercury in glass thermometers are often used in the laboratory but are rarely used on motor vehicles because of their obvious disadvantages. Modern vehicles use electrical thermometers of the type shown in Fig. 2.2.2.

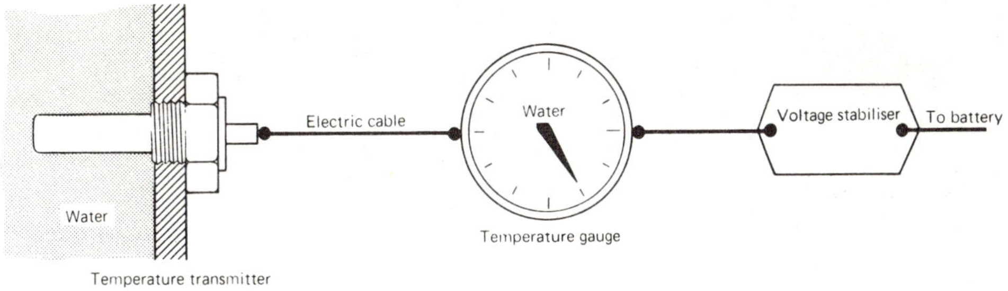

Fig. 2.2.1

2.3 Temperature scales

Most temperature scales are based on two fixed points given by the temperature of

(a) steam from water boiling under standard atmospheric pressure of 760 mm of mercury

(b) pure melting ice.

Celsius scale

Everyday temperature measurement uses this scale which is similar to that introduced in 1742 by the Swedish scientist Andreas Celsius. In this case the freezing point of water is marked 0°C and the boiling point 100°C; temperatures below zero are given negative values. For many years this scale was known as Centigrade but nowadays it is recommended that the name Celsius be used.

Fahrenheit scale

A scale used in Britain until fairly recently. The upper fixed point was marked 212°F and the lower point as 32°F; the interval between divided into 180 degrees. To convert °F to °C the formula is

$$C = (F - 32) \times \frac{5}{9}$$

The SI unit of temperature is the 'kelvin'.

Kelvin scale

This has the same size of degree or temperature interval as the Celsius scale, but has a zero about −273°C. To convert from Celsius to kelvin add 273, e.g.

$$16°C = 16 + 273$$
$$= 289\ K$$

The reader will note that the degree symbol is not used with the kelvin scale.

Value 273 was obtained by noting the behaviour of a gas and this indicated that heat ceased to exist at that temperature. A temperature of −273°C (zero K) is sometimes called the absolute zero.

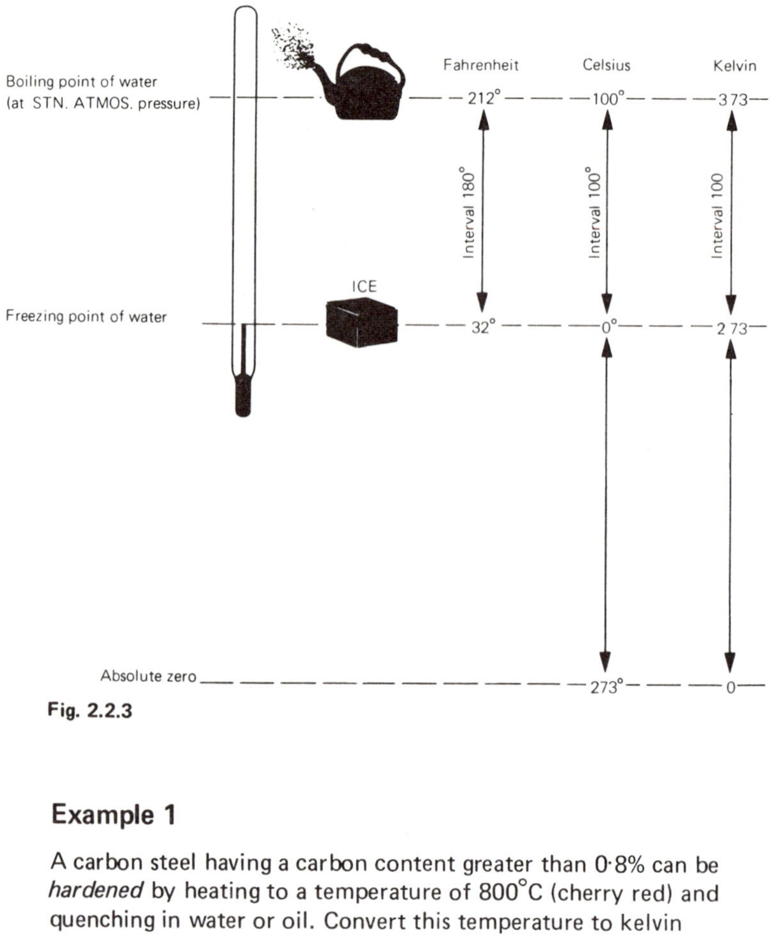

Fig. 2.2.3

Example 1

A carbon steel having a carbon content greater than 0·8% can be *hardened* by heating to a temperature of 800°C (cherry red) and quenching in water or oil. Convert this temperature to kelvin

$$800 + 273 = 1073\ K$$

Example 2

After hardening a steel chisel the brittleness can be relieved by *tempering.* This is performed by heating to a temperature of 573 K (violet colour) and quenching in oil or water.

Convert this temperature to Celsius

$$573 - 273 = 300°C$$

2.4 Heat transfer

Heat is transferred from one substance to another by one or more of the following ways:

Conduction
Convection
Radiation

The actual process is complex so we only need to deal with the effects and application of each.

Conduction

Heat transfer which takes place within the material.

Human beings discover at an early age that heat travels along a metal object. The heat given to the object is passed from particle to particle until the whole thing is at a uniform temperature.

Heat conductivity varies with materials

Some materials such as metals are good conductors of heat whereas materials like asbestos resist the passage of heat and would be called either a bad heat conductor or a good heat insulator. If the heat of an object is to be retained it must be lagged (wrapped) with a heat insulation material. On the other hand, if a large heat 'flow' is required, a material having a good thermal conductivity should be used. Two examples of this are:

(a) *aluminium alloy pistons* – compared to a cast iron, aluminium alloy has a much higher thermal conductivity so this type of piston operates at a lower temperature,

(b) *copper soldering iron bit* – copper is an exceptionally good heat conductor so heat is readily transferred to the material being soldered.

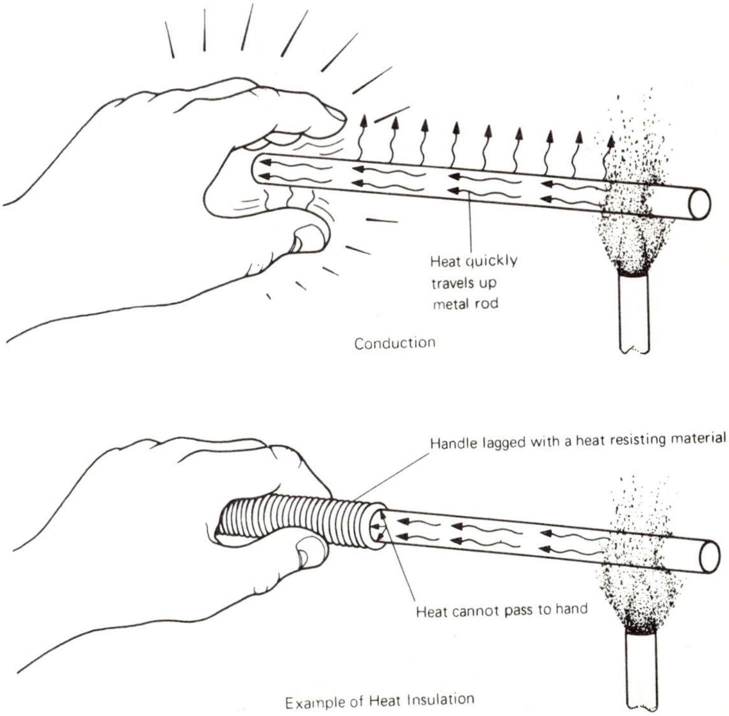

Fig. 2.4.1

Material	Thermal conductivity
Copper	Excellent
Aluminium	Very good
Iron	Good
Lead	Fair
Wood	Poor
Asbestos	Very poor

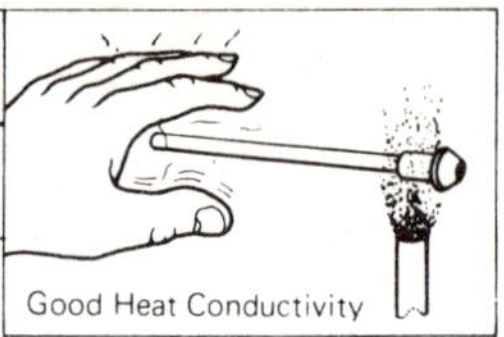

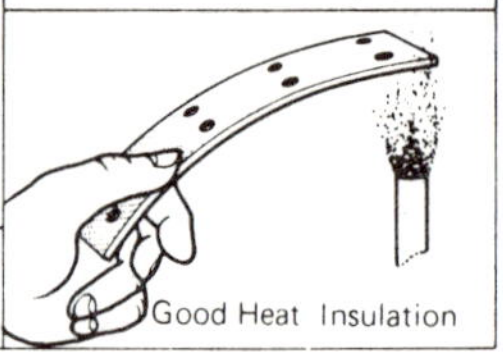

Comparison of Thermal Conductivity of Materials

Fig. 2.4.2

Liquids such as water, petrol and oil are poor conductors of heat and gases are even poorer. Liquids and gases normally transfer their heat by convection which we will consider next.

Convection

Heat transfer due to the movement of the fluid

Most readers will have seen cases where fluid has been heated and movement of the fluid has taken place. Experiments similar to Fig. 2.4.3 show that when water at a temperature above 4°C is heated, it expands, decreases its density and rises towards the surface.

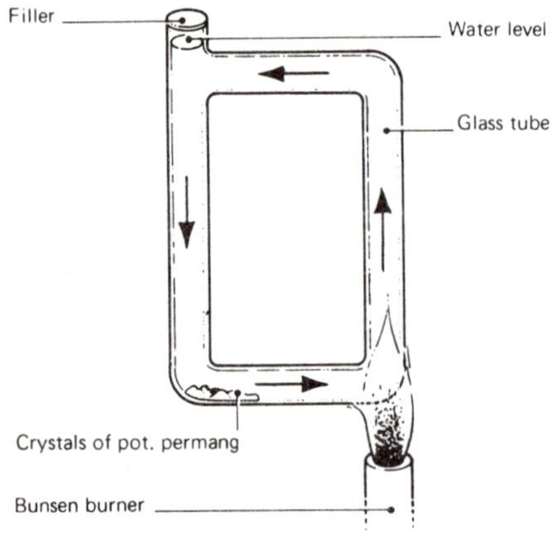

Fig. 2.4.3

Water gets hotter → expands → gets lighter → rises

Similar effects can be demonstrated by using other liquids and gases, e.g. air.

Movement of water by convection can be employed to operate a simple engine cooling system (see F of M.V.T.).

Radiation

Heat transfer by rays which are similar to light rays

One of the best examples of this form of heat transfer is the sun. Although the distance between the sun and earth is great, the sun's rays are powerful and are capable of travelling through space. In a

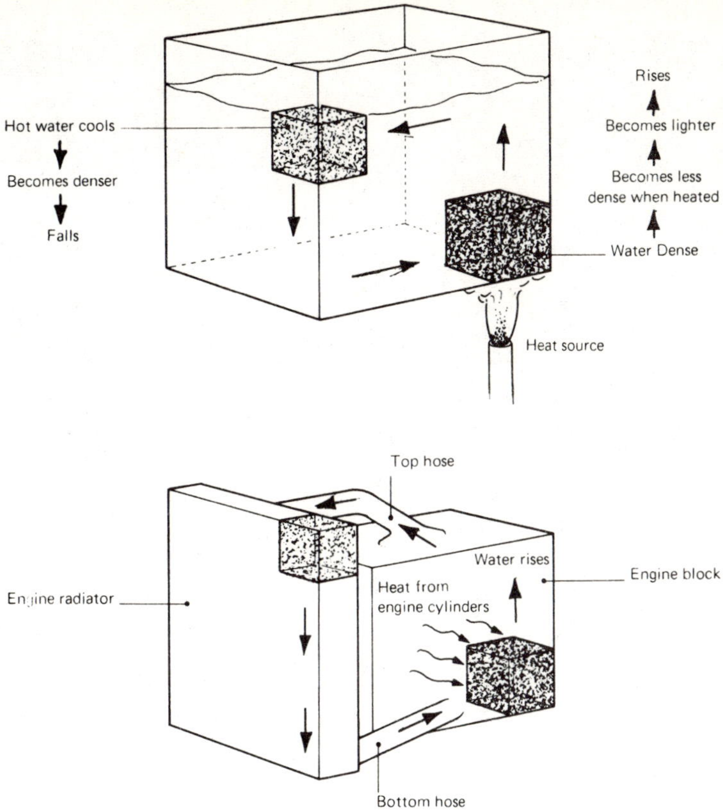

Water Movement By Convection

Fig. 2.4.4

similar way, the heat from any source can be felt when there is no interruption to the path of the rays. This radiant heat consists of invisible electromagnetic waves which behave in a manner similar to light waves:

(i) they travel in a straight line
(ii) they can be reflected as shown in Fig. 2.4.5.

Rate of heat transfer by radiation depends on:

(i) temperature of the source
(ii) area of its surface
(iii) nature of its surface.

Obviously the hotter the source the greater will be the transfer.

Increasing the area has the effect of 'spreading out' this hot surface and allows the heat to be given up more readily.

The nature of the surface is the finish; a dull black finish radiates heat better and also absorbs heat better than a highly polished surface.

M.V. applications of radiation

(a) *Fins around an air-cooled engine cylinder* increase the surface area exposed to the air (Fig. 2.4.5).

(b) *Radiator of a liquid cooling system.* A large area of hot water is presented to the air.

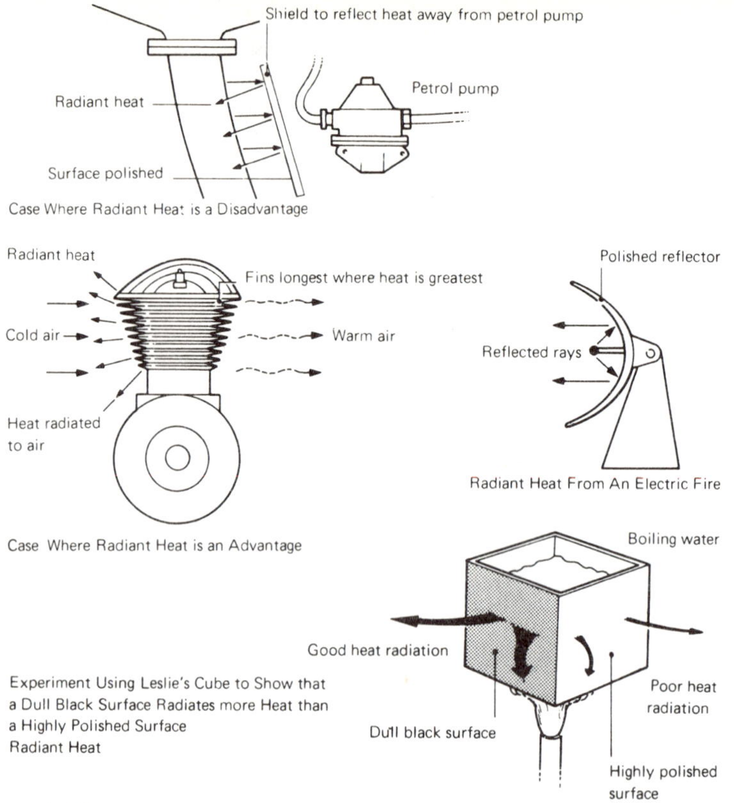

Fig. 2.4.5

In these cases the rate of flow will depend on the difference in temperature between the hot surface and the cold air.

2.5 Thermal expansion and contraction

Expansion of solids

The fact that most substances increase in size on heating and decrease in size on cooling has already been mentioned. The general effect is the same for all substances, but there are important differences in detail between what happens in solids, in liquids and in gases.

Heating a solid causes it to expand in all directions but one of the most important considerations is its increase in length or linear expansion.

Linear expansion is the expansion in length

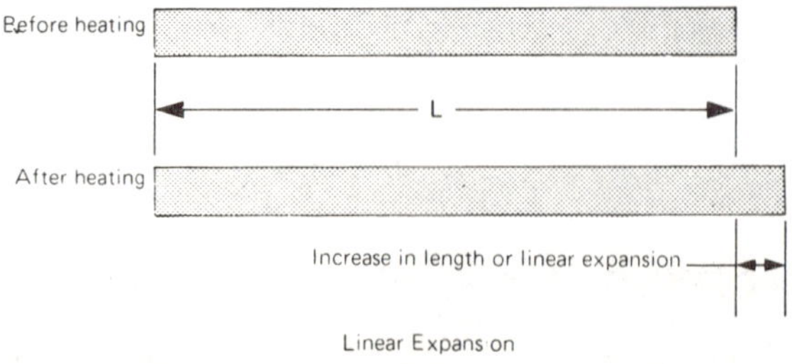

Fig. 2.5.1

This increase in length must be taken into account when the mechanic assembles a component, since failure to observe this fact may result in considerable damage to the unit. The most common way to prevent damage is to consider the amount that the item will expand and then allow a clearance for this to take place. There are a number of interesting cases in motor vehicles where linear expansion is taken into account by giving a clearance.

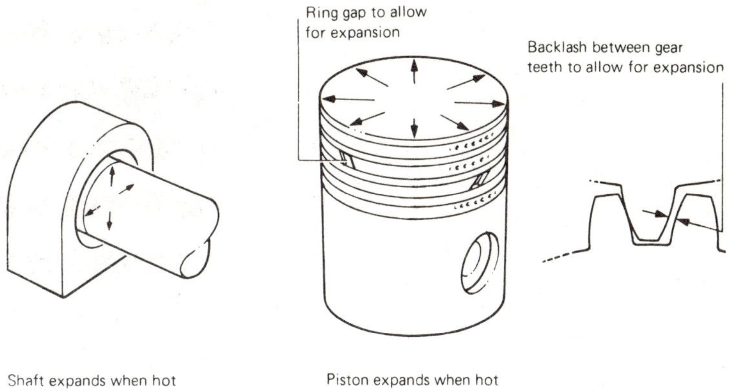

Expansion of M.V. Components

Fig. 2.5.2

(a) *Shafts running in bearings* – when the shaft gets hots its diameter increases, so if the bearing did not expand the same amount, seizure would occur.

(b) *Piston in a cylinder* – both get hot but the piston generally expands more than the cylinder. An aluminium piston of diameter 80 mm expands about 0·5 mm when it is heated to its working temperature.

(c) *Gear teeth* – on heating the teeth get bigger, so a backlash (clearance between gears) is provided.

All materials do not expand at the same rate

Experiments show that the linear expansion varies with different materials and Fig. 2.5.3 shows the variation for a given temperature use.

The table indicates that aluminium expands more than twice as much as cast iron and this shows why a piston made of aluminium alloy must be given a larger clearance than a similar piston of cast iron. In the case of pistons, expansion is a nuisance so in order to control it, the designer may either restrict the heat flow or incorporate in the piston a 'strut' to resist the expansion. A material for such a 'strut' is Invar.

The bi-metal strip

We have seen that brass and steel expand at different rates so if these two materials are joined together the effect of heating will be to bend the strip (Fig. 2.5.4). This principle is often used to operate various devices fitted to motor vehicles, e.g. electrical contacts in instruments.

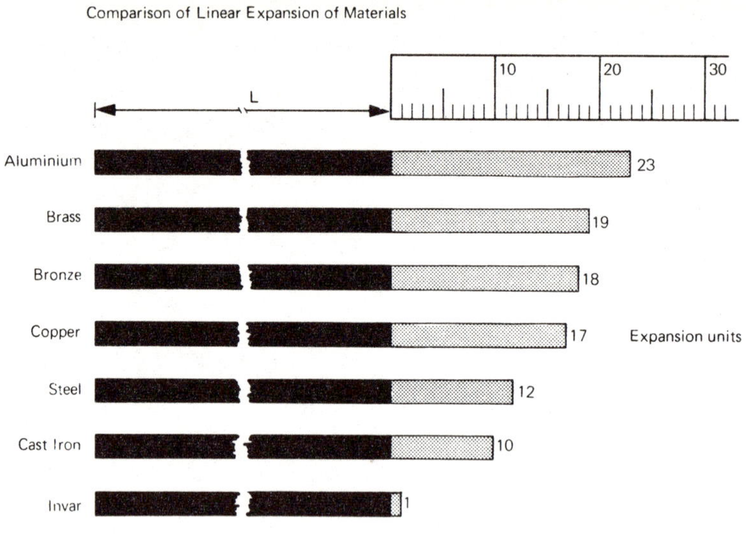

Fig. 2.5.3

Linear expansion can be utilised by the mechanic to aid him when assembling or dismantling units.

M.V. applications of expansion of solids

(a) *Flywheel starter ring gear* – a tight fit must be provided between gear and flywheel. The gear is heated until the diameter is large enough to accept the flywheel (Fig. 2.5.5).

(b) *Gudgeon pin in piston* – if the pin is tight to remove, the piston can be expanded in hot oil – the pin can be easily removed because the piston expands more than the pin (Fig. 2.5.6).

(c) *Seized pins* – there are occasions when a pin cannot be removed and heat is applied to the female part. Special consideration must be given to the effect of the heat on the material before using this method.

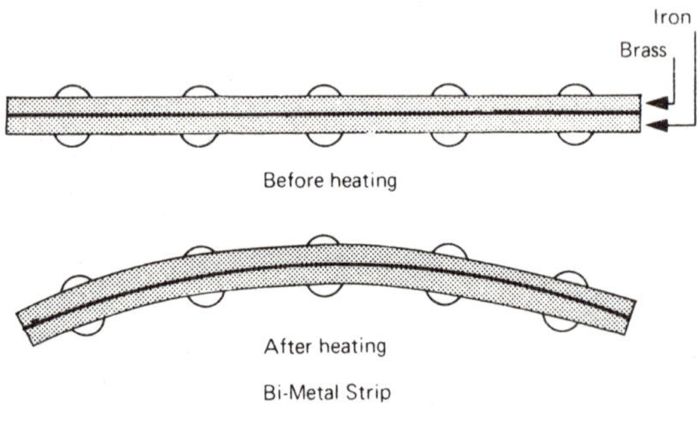

Fig. 2.5.4

All the above examples are based on expansion – sometimes the material is made to contract, e.g.:

Valve guides and valve inserts – sometimes these are fitted after they have been contracted by immersion in a liquid gas such as carbon dioxide (Fig. 2.5.7). This gas changes to liquid at a temperature of 195 K (–78°C).

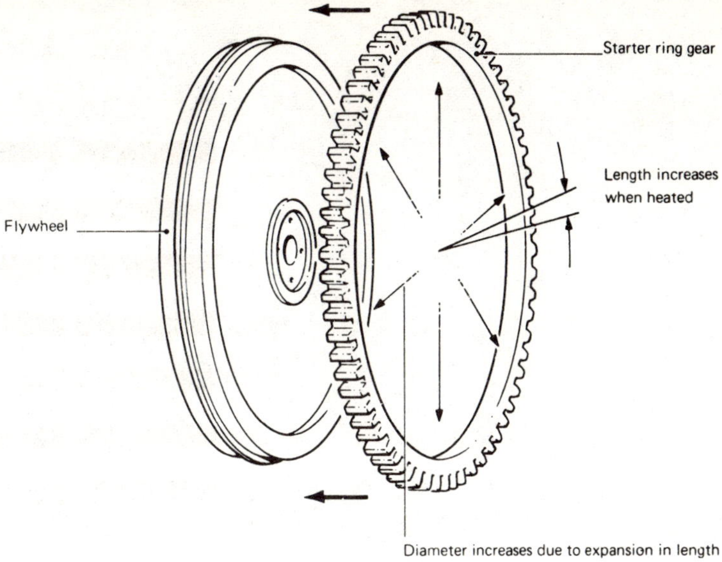

Fitting a Starter Ring Gear to Flywheel

Fig. 2.5.5

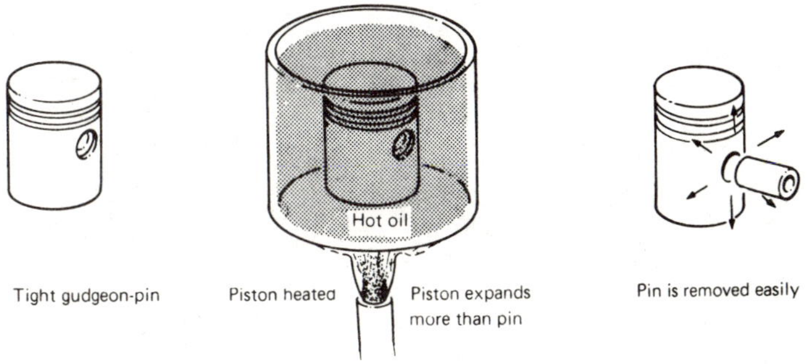

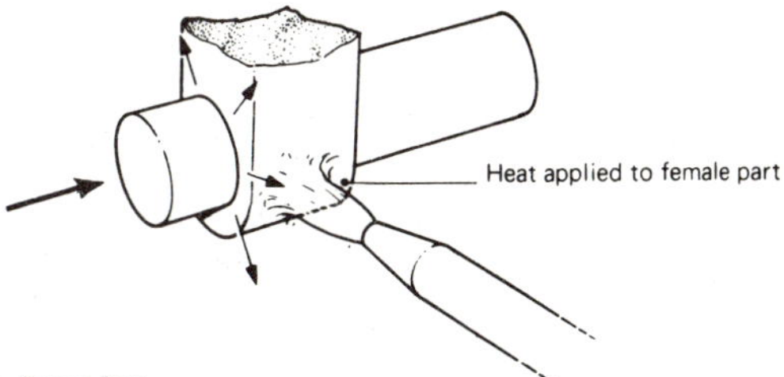

Removal of Tight or Seized Pins

Fig. 2.5.6

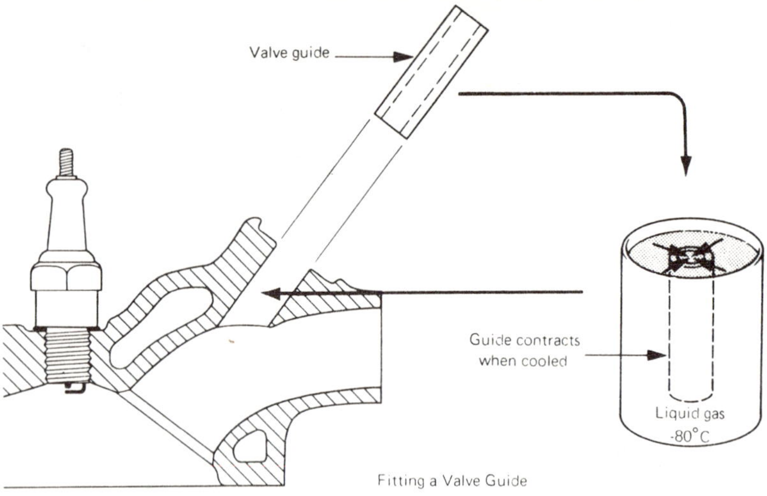

Fig. 2.5.7

Increase of area and volume

As stated already, materials expand in all directions; the increase in:

area is called superficial expansion,
volume is called volumetric expansion.

Expansion of liquids

Owing to their nature, the expansion of liquids must be treated on a volume basis. As shown by the mercury in a thermometer, most liquids increase their volume as they are heated, but the real amount they expand is difficult to demonstrate since it must take into account the expansion of the container. If the container did not alter its size then the diagram (Fig. 2.5.8) would indicate the *real volumetric expansion of* various liquids. When the container expansion is not considered the term used is *apparent volumetric expansion.* It has been stated that most liquids expand on heating – this means that there are exceptions and water is one of these special cases. Above 4°C (277 K) water expands on heating but if the temperature is lowered below 4°C the water also expands. This means that at 4°C the water is at its greatest density. When the temperature is lowered to 0°C water changes to ice and this is accompanied by a considerable increase in volume.

The expansion due to the change of state from water to ice can cause considerable damage to a liquid cooled engine. Large forces, sufficient to crack a cylinder block, occur at the change, so to prevent this possibility, the risk of freezing must be reduced. This is achieved by adding to the water a liquid such as ethylene glycol. As the quantity of 'anti-freeze' in the water is increased so the temperature of freezing is decreased. (See F of M.V.T.)

Many liquids are used in motor vehicles and if any of these are heated, an increase in their volume will occur.

M.V. applications of expansion of liquids

(a) *Cooling system* – an overflow is provided to allow for expansion. Some systems provide a separate container to collect this discharge so as to enable it to be returned when the system has cooled.

(b) *Braking system* – an air vent and expansion space is provided in the reservoir. If the reservoir was completely filled with brake

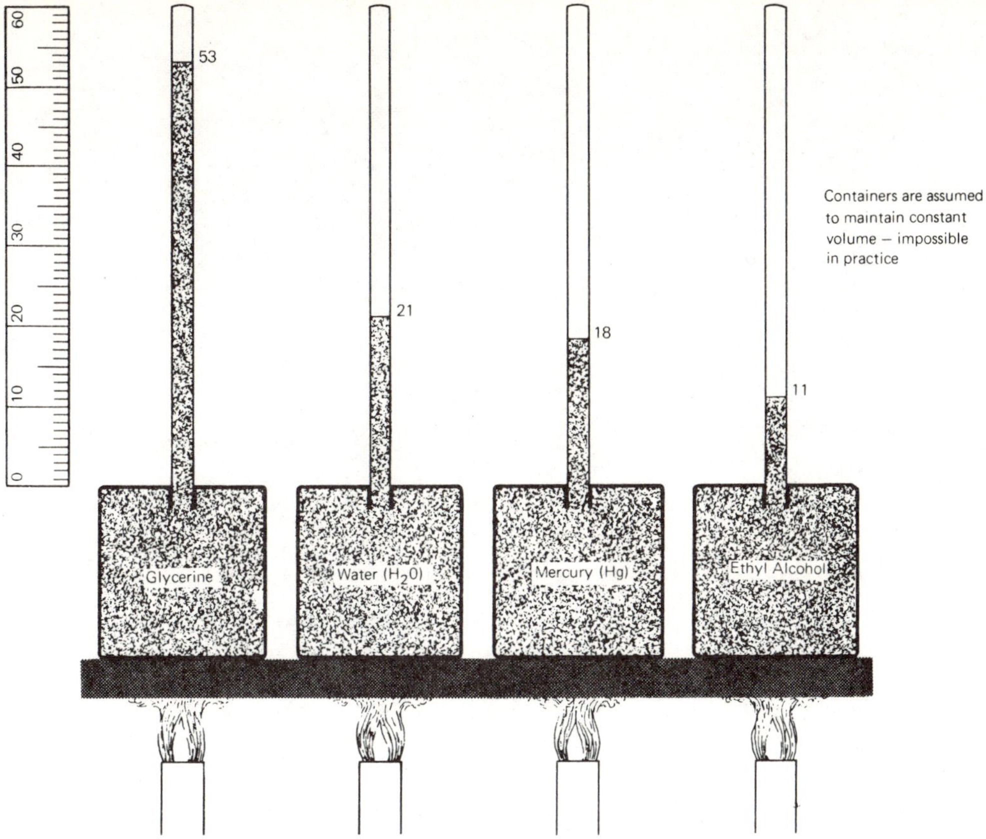

The Real Expansion of a Liquid for a Given Temperature Rise

Fig. 2.5.8

fluid when the system was cold, then the system would overflow after the brakes had been used.

(c) *Lubricating oil in steering box, gearbox, final drive etc.* – unless a breather was provided, expansion of the oil would build up a pressure which could force the oil past the oil sealing arrangements.

Expansion of gases

Solids and liquids are almost incompressible and there is very little difficulty in determining their rate of expansion. Gases, however, can easily be compressed, so in order to determine the expansion due to heating great care must be taken to ensure that the pressure of the gas is not altered.

Consider a given volume of gas contained in the cylinder shown in Fig. 2.5.9 which is fitted with gas-tight piston and loaded with a mass to maintain a constant gas pressure. During heating the gas will expand and this will cause the piston to move. So:

the volume increases as the temperature increases, if the pressure is kept constant.

By conducting a similar experiment with the piston locked to form a closed cylinder (Fig. 2.5.10) the effect of a constant volume can be examined. In this case the gas is prevented from expanding,

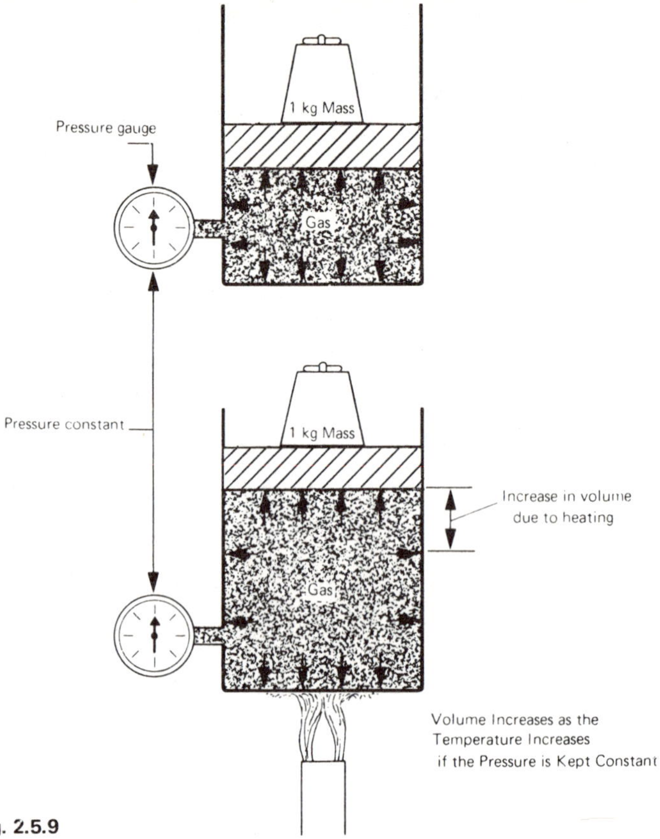

Fig. 2.5.9

therefore it exerts a pressure which gets higher and higher as the temperature increases. So:

the pressure increases as the temperature increases if the volume is kept constant

M.V. applications of expansion of a gas

(a) *Gas in an engine cylinder.* Ignition of the fuel causes a rapid rise in temperature while the piston is positioned about top dead centre, so there is practically no change in volume of the gas during this burning process. As a result of this temperature increase, a considerable rise in pressure occurs and it is this pressure which provides the force on the piston to drive the engine.

(*b*) *Air in a tyre.* The rotation of a wheel causes heat to be generated which raises the temperature of the tyre. The volume remains about the same, so the pressure will increase. To allow for this, the manufacturers normally state that tyres should be checked when 'cold'.

Besides the effect of altering the temperature of a gas we must also see what happens to the pressure when only the volume is altered. This can be visualised when a gas is trapped in a cylinder (Fig. 2.5.11) and the volume is reduced. The graph shows that if the volume is halved the pressure is doubled. So:

the pressure increases as the volume decreases, if the temperature is kept constant

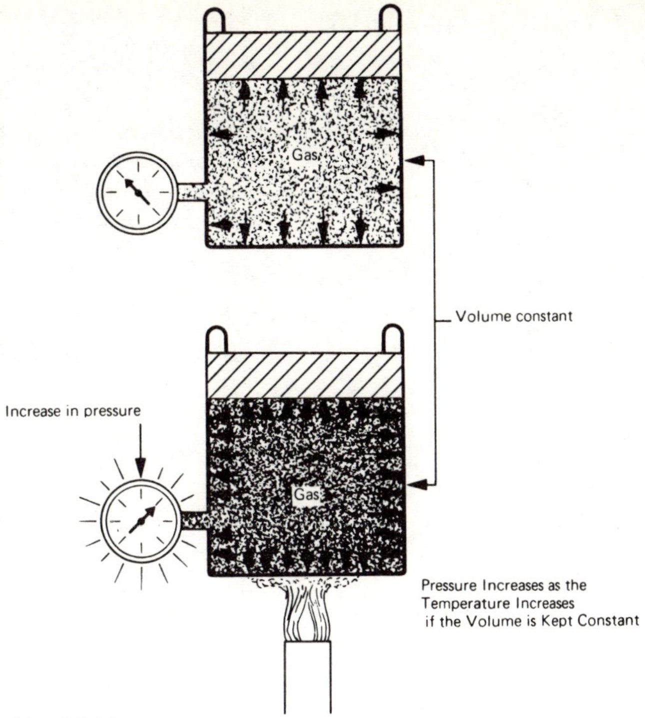

Fig. 2.5.10

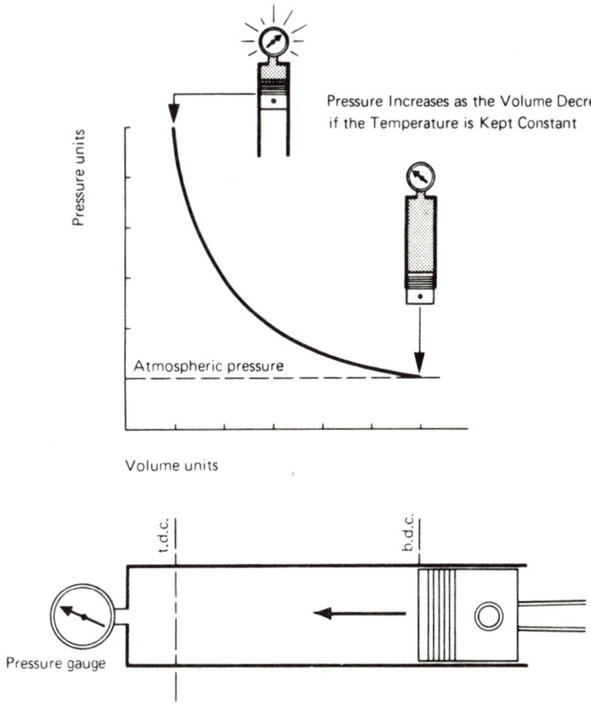

Fig. 2.5.11

2.6 Change of state

Materials exist in one of three states:

solid, liquid or gas

Heat can change a material from one state to another (Fig. 2.6.1).

The process by which a solid turns into a liquid is called *fusion* or *melting*, and the reverse process by which a liquid turns into a solid is called *solidification* or *freezing.*

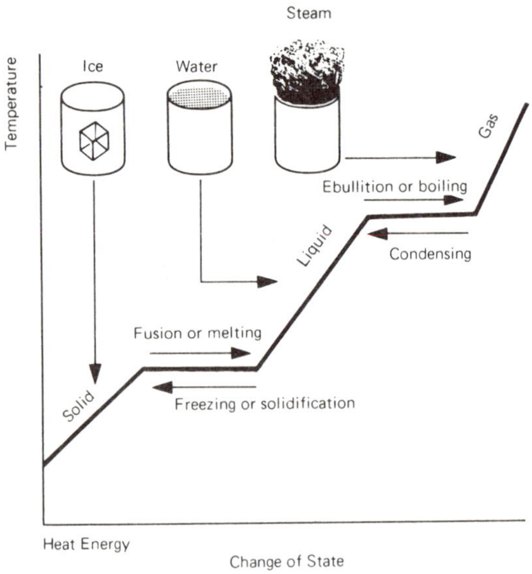

Fig. 2.6.1

The change from liquid to vapour is called vapourisation: this may take place relatively slowly and from the surface of the liquid only, in which case it is called *evaporation*: or it may take place rapidly with bubbles of vapour forming within the bulk of the liquid itself and rising to the surface, in which case it is known as *ebullition* or *boiling.* The reverse process by which a vapour changes into a liquid is called *condensation.* These changes generally take place at definite temperatures in any particular material, but at different temperatures in different materials.

Pure water freezes at a temperature of 0°C (273 K) when exposed to normal atmospheric pressure, and during the process of freezing its volume increases slightly. This expansion on solidification is shared by some other materials (e.g. cast iron, in which the expansion on solidification helps in obtaining accurate castings) but not by all. Paraffin wax is an example of a material which shrinks on solidification (Fig. 2.6.2) and conversely expands on melting – hence its use in one type of thermostat (see F. of M.V.T.).

Pressure and boiling point

All liquids tend to form vapour in the space above the surface. Some vapourise more readily than others and petrol and alcohol are good examples – these are said to be volatile.

Consider a case where the space above a liquid is enclosed. Any vapour which escapes from the liquid will create a pressure in the air

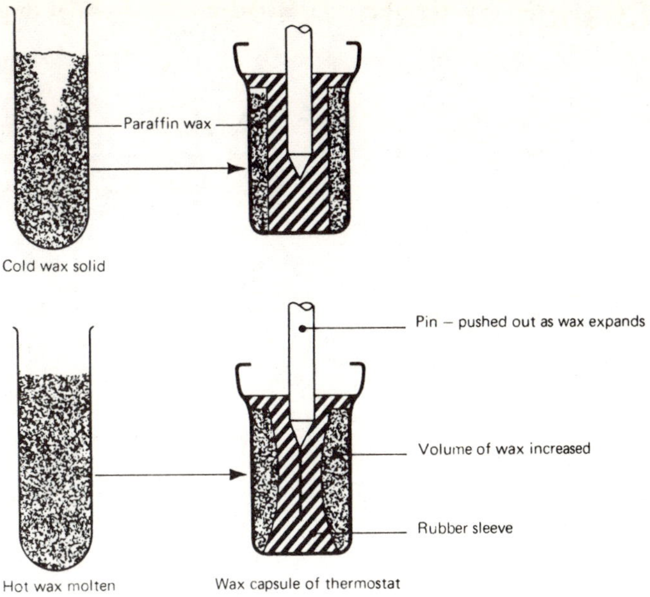

Effect of Heat on Wax

Fig. 2.6.2

space which will prevent any more vapour from leaving the liquid (Fig. 2.6.3).

If the temperature of the liquid is raised, the greater internal pressure, i.e. vapour pressure, will enable more vapour to escape from the liquid. In turn this vapour will increase the external pressure, and prevent further escape of vapour.

By repeating this process a temperature is reached when the internal pressure becomes equal to the external pressure. At this point the vapour leaves the liquid in a vigorous manner and the liquid is said to boil.

A liquid boils when the vapour pressure equals the external pressure

When the lid of the container is removed, the external pressure is that of the atmosphere so if this external pressure is increased the previous example shows that the liquid must reach a higher temperature to boil.

Similarly if the atmosphere pressure is lowered, the vapour pressure to equalise this will be reached at a lower temperature, i.e. the liquid will boil at a lower temperature. So:

the boiling point of a liquid depends on the pressure

M.V. applications of pressure and boiling point

(a) *Cooling system* – the heat flow from the combustion chamber to the coolant depends on the temperature difference so the coolant should be kept as hot as possible. This is difficult with a cooling system which is open to the atmosphere because if it is

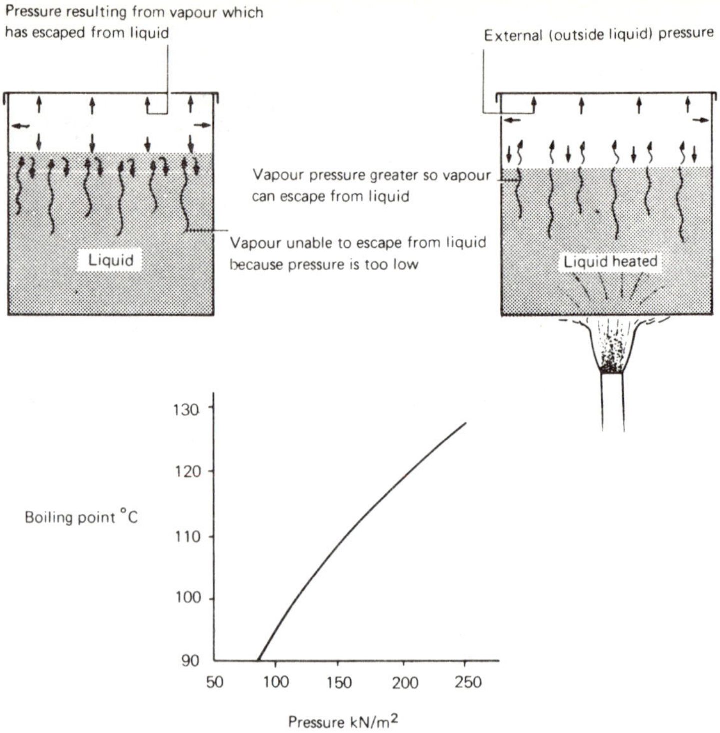

Effect of Pressure on Boiling Point of Water

Vapour Pressure & Boiling Point

Fig. 2.6.3

operated in excess of about 90°C there is a risk of the coolant boiling. By pressurising the system the operating temperature can be raised and the engine become more efficient (see F of M.V.T.).

(b) *Fuel lines* – the petrol in the pipe between the tank and the pump is at a pressure less than atmospheric. This low pressure will allow the petrol to give off more vapour and may result in a reduced petrol supply, or on a hot summer's day, no supply whatsoever. To prevent this trouble the pipe from the tank to the pump should be as short as possible.

BEFORE ATTEMPTING EXERCISES, REFER TO PAGE 175 WHICH GIVES SUGGESTIONS ABOUT THE TECHNIQUE OF ANSWERING MULTIPLE CHOICE TYPE QUESTIONS.

Exercises 2.7

Each question shows four possible answers. Select which is correct.

1. What is meant by the term 'heat'?
(a) The hotness of a body
(b) The degree of hotness
(c) Internal energy capable of doing work
(d) Work which is stored in a body

2. In which direction does heat flow?
(a) Upwards
(b) Downwards
(c) From a hot to a cold object
(d) From a cold to a hot object

3. The temperature of a body is the
(a) heat contained in that body
(b) degree of hotness
(c) change of state
(d) rate of conduction

4. Temperature is measured by a
(a) hydrometer
(b) thermometer
(c) galvanometer
(d) micrometer

5. What happens to mercury when it is heated?
(a) It contracts
(b) It expands
(c) It changes the shape of the container
(d) It vapourises

6. A temperature of 6° Celcius is the same as
(a) 6° Fahrenheit
(b) 6° Centigrade
(c) 6 Kelvin
(d) 6 mm mercury

7. Under standard conditions the freezing point and boiling point of pure water are
(a) 32°C and 212°C
(b) 0°C and 212°C
(c) 32°C and 100°C
(d) 0°C and 100°C

8. Under standard conditions the freezing point and boiling point of pure water are
(a) 0 K and 100 K
(b) 0 K and 373 K
(c) 273 K and 373 K
(d) 305 K and 485 K

9. It is considered that heat ceases to exist at a temperature of absolute zero. This occurs at
(a) 0 K
(b) 273 K
(c) 0°C
(d) 32°C

10. What is the SI unit of temperature? The
(a) centigrade
(b) fahrenheit
(c) celsius
(d) kelvin

11. Which one of the following is a typical temperature for hardening a high carbon steel?
(a) 100°C
(b) 400°C
(c) 800°C
(d) 1200°C

12. Which one of the following is a typical temperature for tempering a chisel?
(a) 50°C
(b) 300°C
(c) 550°C
(d) 800°C

13. State the method of transferring the heat from the piston crown to the piston skirt.
(a) Expansion
(b) Conduction
(c) Convection
(d) Radiation

14. State the method of transferring the heat from an electric lamp filament to the glass bulb.
(a) Expansion (c) Convection
(b) Conduction (d) Radiation

15. State the method of transferring the heat from the cylinder block to the radiator
(a) Expansion (c) Convection
(b) Conduction (d) Radiation

16. Which one of the following materials has the highest heat con-conductivity?
(a) Asbestos (c) Lead
(b) Aluminium (d) Steel

17. To prevent loss of heat a body should be
(a) wrapped with a cold material
(b) wrapped with a material having a high thermal conductivity
(c) lagged with a heat insulation material
(d) lagged with an aluminium foil

18. Which one of the following is the best conductor of heat: steel, water, oil, air?
(a) Steel (c) Oil
(b) Water (d) Air

19. What causes convection in an engine cooling system?
(a) As water is heated it expands, gets lighter and rises
(b) As water is cooled it expands, gets heavier and falls
(c) The pressure given by the pump
(d) The pressure drop at the narrow top hose

20. What type of heat can be reflected?
(a) Conducted (c) Radiant
(b) Convected (d) Light

21. What is the purpose of the fins around the cylinder of a single cylinder engine?
(a) To control expansion (c) To radiate the heat
(b) To strengthen the cylinder (d) To reduce the noise

22. Which one of the following radiates the greatest amount of heat?
(a) A hot surface having a large area
(b) A cold surface having a small area
(c) A cold surface having a large area
(d) A hot surface having a small area

23. What is meant by the term 'linear expansion'?
(a) The heat required to produce expansion
(b) The expansion of any straight object
(c) The increase in length due to expansion
(d) The gradual expansion of aluminium

24. What provision is made to allow for thermal expansion of a piston ring?
(a) A gap clearance is given

(b) The ring is kept in contact with the cylinder wall
(c) The ring is pegged
(d) The ends of ring are butted together

25. Which one of the following is a suitable method for fitting a starter ring gear onto a flywheel?
(a) The gear is cooled and forced into the flywheel
(b) The flywheel is heated and forced into the gear
(c) The gear is heated until it is larger than the flywheel
(d) The flywheel is cooled until it is smaller than the gear

26. Which one of the following is a suitable method of removing a tight gudgeon pin from a piston?
(a) The pin is cooled by immersion in air at temperature of −78°C
(b) The pin is cooled by immersion in air at temperature of 195 K.
(c) The piston is expanded by immersion in water at temperature of 110°C
(d) The piston is expanded by immersion in oil at temperature of 110°C

27. When a brass-steel bi-metal strip is heated the effect is that the strip
(a) lengthens equally on both sides
(b) twists in a clockwise direction
(c) bends in a direction away from the brass
(d) bends in direction towards the brass

28. Which one of the following has the lowest rate of linear expansion
(a) Aluminium
(b) Brass
(c) Cast iron
(d) Invar steel

29. Which one of the following has the highest rate of linear expansion
(a) Aluminium
(b) Brass
(c) Cast iron
(d) Invar steel

30. Which two words make the following statement correct?
An aluminium alloy piston expands about . . . . . as much as a similar cast iron piston so a . . . . skirt clearance is necessary with the lighter piston
(a) Half . . . . smaller
(b) half . . . greater
(c) twice . . . . smaller
(d) twice . . . . greater

31. The increase in volume of a liquid heated in a container which does not alter its shape is an indication of the liquid's
(a) apparent volumetric expansion
(b) real volumetric expansion
(c) superficial expansion
(d) superficial volumetric expansion

32. The greatest density of water occurs at
(a) −2°C
(b) 0°C
(c) 4°C
(d) 40°C

33. Anti-freeze is added to a cooling system to
(a) prevent the system freezing

(b) prevent ice forming at temperatures below 0°C
(c) lower the temperature at which freezing occurs
(d) control the increase in volume when water changes to ice

34. A cracked cylinder block has resulted from freezing of the coolant. The crack occurs when the
(a) water changes its state
(b) ice changes its state
(c) ice changes to water
(d) water changes to steam

35. Which two words make the following statement correct?
The pressure of a gas is kept constant, so a temperature . . . . will cause a volume . . . .
(a) rise . . . . fall
(b) fall . . . . rise
(c) decrease . . . . increase
(d) increase . . . . increase

36. Which two words make the following statement correct?
The volume of a gas is kept constant, so a temperature . . . . will cause a pressure . . . .
(a) rise . . . . fall
(b) fall . . . . rise
(c) decrease . . . . increase
(d) increase . . . . increase

37. Which two words make the following statement correct?
The temperature of a gas is kept constant, so a volume . . . . will cause a pressure . . . .
(a) rise . . . . rise
(b) fall . . . . fall
(c) decrease . . . . increase
(d) increase . . . . increase

38. A liquid boils when the vapour pressure equals the
(a) external volume
(b) external temperature
(c) internal pressure
(d) external pressure

39. When a cooling system is pressurised the effect is to
(a) lower the boiling point
(b) raise the boiling point
(c) allow the engine to heat up quicker
(d) allow the engine to heat up slower

40. Which two words make the following statement correct?
When the petrol in the fuel line is at a pressure . . . . than atmospheric it will vapourise at a . . . . temperature
(a) lower . . . . constant
(b) higher . . . . constant
(c) lower . . . . lower
(d) higher . . . . lower

2 SCIENCE (b) Mechanics

2.10 Force

We are all familiar with the fact that things have to be pulled or pushed to make them move. *Pulls* and *pushes* are examples of what we call *force.* This we cannot see or touch but we are aware of its existence from its effects, so from these effects we can say:

force is that which changes a body's state of rest or of uniform motion in a straight line

This statement is a technical expression of facts which are generally known. Translating it more into simple words:

(i) If an object is stationary it will remain stationary. A force must be applied to move it.

(ii) When an object is moving then it will continue to move in the given direction and speed unless a force is applied to it.

In order to investigate this subject more fully we must consider some more basic terms. Firstly, let us look at the object or body onto which the force is acting. The quantity of matter in the object is its *mass* and this would be expressed in *kilogrammes.*

Assuming that the object or mass is free to move, the application of a force will cause the object to move in the direction that the force is acting. At first the object will move slowly, but all the time the force acts the velocity of the object will increase.

The rate of increase in velocity is called acceleration.

Let us assume that the velocity increases by 2 metre/second for every second that the force is acting:

After 1 second the velocity would be 2 metres/second
After 2 seconds the velocity would be 4 metres/second
After 3 seconds the velocity would be 6 metres/second
After 4 seconds the velocity would be 8 metres/second.

In each second of time, the velocity has increased by 2 m/s, so the acceleration is said to be 2 metres per second, per second, which in symbol form is 2 m/s/s or more correctly stated as 2 m/s^2.

Now let us return to the force which was accelerating the object and consider the unit of force. SI adopts a unit which is directly linked (coherent) to the base units and to honour the work of Sir Isaac Newton

the SI unit of force is the newton and given the symbol N.

(You will note that when the name of a person is used as a unit, small letters are used for the name of the unit and a capital letter used for the symbol.)

One newton is the force which, acting on a mass of one kilogramme, gives to that mass an acceleration of one metre per second per second.

$$1 \text{ newton} = 1 \text{ kg} \times 1 \text{ m/s}^2$$

Weight

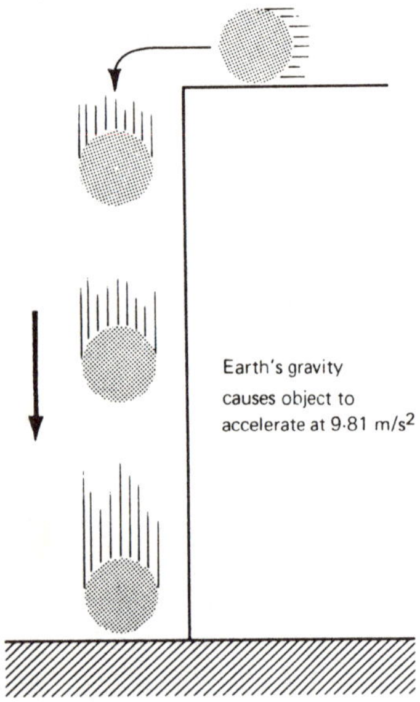

Fig. 2.10.1

The force with which we are most familiar is that produced by the earth's gravity since any object on or near the surface of the earth is attracted towards the centre of the earth. This force of gravity is proportional to the amount of material the object contains; in other words its mass in kilograms. In everyday language the force of gravity on a mass is called weight but this term has technical limitations due to the :

(i) variation in gravitational attraction in different parts of the earth's surface.

(ii) lack of gravity when the object is away from the earth, i.e. in space.

These limitations show that the weight of an object depends on its location, so to overcome these drawbacks the term mass is preferred: this will remain the same irrespective of the location of the object.

In various problems given in this book the term 'mass' has been used to indicate the 'size' of an object. However the reader must expect to see the word 'weight' applied to many vehicle applications since apart from a few exceptions it is expected that the normal motor vehicle will operate within the earth's gravitational field.

Accepting the fact that any large mass produces a gravitational force, then it follows that if an object is dropped from the top of a tall building, it will fall towards the ground at a velocity which increases with time, i.e. the mass will accelerate due to the force of gravity. Taking accepted values for an object falling in a vacuum:

g or gravitational acceleration 9·81 m/s².

This acceleration applies to objects big and small — they all accelerate at the same rate.

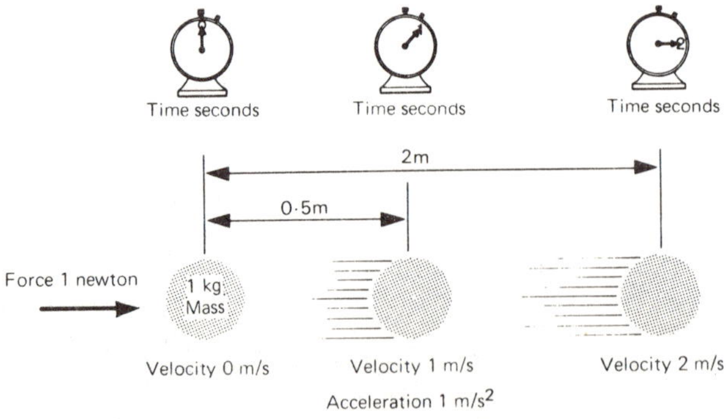

Fig. 2.10.2

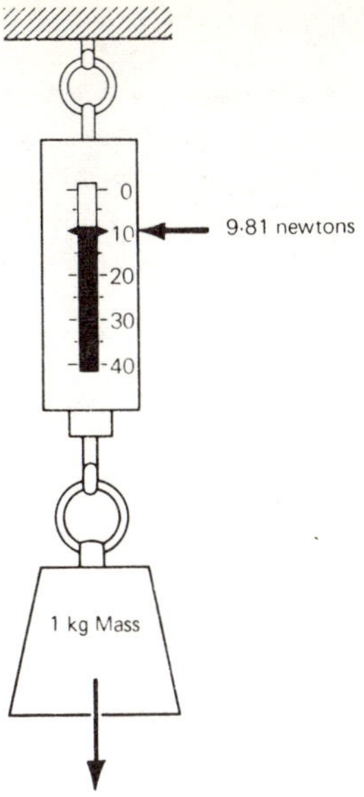

Mass Attracted by Earth's Gravity

Fig. 2.10.3

Referring to the definition of a newton which is:

1 newton is required to accelerate a mass of 1 kg at a rate of 1 1 m/s^2 then it is possible to state that:

9·81 N is required to accelerate a mass of 1 kg at a rate of 9·81 m/s^2.

The effect of the earth's gravity causes a mass of 1 kg to give a force of about 9·81 N

i.e. Force of gravity on mass of 1 kg $\hat{=}$ 10 N

Example 1

A 4 wheeled vehicle has a mass of 2000 kg. What force acts between the tyre and road assuming the force on each wheel is equal?

$$\text{Mass acting on each wheel} = \frac{2000}{4} = 500 \text{ kg}$$

1 kg gives a force of 9·81 N

$$\text{so 500 kg} \quad = 500 \times 9{\cdot}81 \left(\text{or } 1000 \times \frac{9{\cdot}81}{2}\right)$$

$$= 4950 \text{ N} = 4{\cdot}905 \text{ kN}$$

(Check – answer should be about 10 times larger than the mass acting on each wheel.)

Example 2

A vehicle positioned on a lift requires a force of 29·43 kN to keep it in the raised position. What mass is acting on the lift?

1 kg mass gives force of 9·81 N

$$\therefore \text{ Mass acting on lift} = \frac{29\,430}{9{\cdot}81} = 3000 \text{ kg or 3 tonne}$$

(Check – answer should be about 10 times smaller than the force.)

Prior to the introduction of SI, the units of mass in Britain was the pound (lb) and the gravitational force on this mass was 'one pound force', which was written as 1 lbf. Using this convention, a mass of 1 kg had a gravitational force of 1 kgf. Introduction of the newton as the SI unit of force has made redundant the units 'lbf' and 'kgf', but it is expected that the reader will encounter these units for some time.

To convert to SI units the following can be used:

1 lbf $\hat{=}$ 4·45 newtons
and 1 kgf $\hat{=}$ 9·81 newtons

2.11 The moment of a force

If we want to make something turn, or rotate, we must support it at the point about which we want it to turn. This is to ensure that it cannot move at this point, but can turn around it.

There are many situations where the mechanic can confirm this fact, e.g. Any attempt to apply a force to a spanner to unscrew the nut shown in Fig. 2.11.1 is not very successful – the whole thing floats away in the direction of the arrows. The mechanic's answer is to clamp the bolt in a vice. Now the force acting on the spanner is resisted by the vice and this provides an equal and opposite force to that exerted by the mechanic.

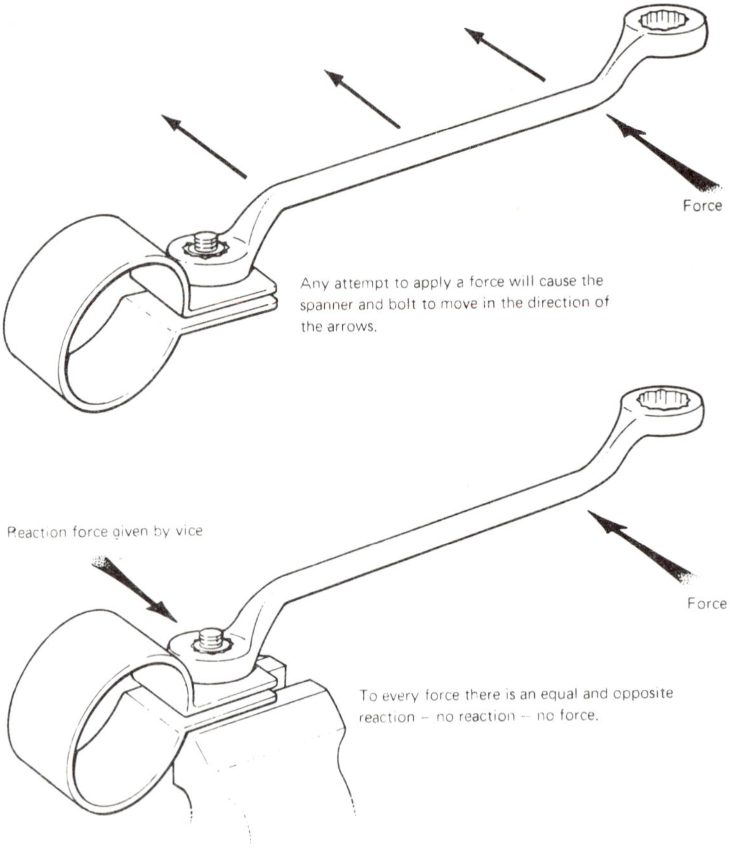

Fig. 2.11.1

This example shows that a force cannot act unless there is a reaction or:–

to every force there is an equal and opposite reaction.

Let us examine some other facts about the use of this spanner. You are all aware that a force applied to a spanner as shown in Fig. 2.11.2a would not have much effect on a tight nut. The line of action of the force given by the mechanic acts through the centre of the nut so no turning effect is obtained. Setting the spanner in the manner shown in Fig. 2.11.2b is not much better but when it is repositioned to enable its full length to be used (Fig. 2.11.2c) the maximum turning effect is obtained and nut can be moved with the minimum of effort. This example shows that the turning effect is governed by two things; the force F and the distance r. The product of F and r is called the *moment of a force* and when this is applied to things that rotate, then terms such as **'turning moment'** or **'torque'**

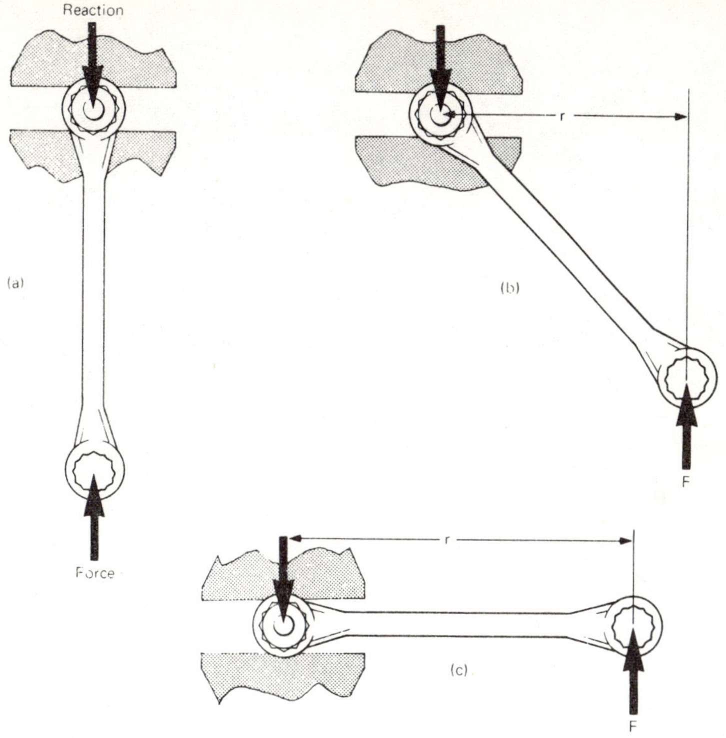

Turning Moment of a Spanner

Fig. 2.11.2

are used. **Torque is a turning moment about a point and is the product of the force and the perpendicular distance between the point and the line of action of the force**

Torque = $F \times r$

Referring to Fig. 2.11.2 shows that when the spanner is positioned at *a* the torque is zero. Moving the spanner towards B causes the distance *r* to increase and this distance is a maximum at *c*. This means that when a constant force is applied, maximum torque is produced in position *c*.

Torque is a product of a force, which is given in newtons, and distance which is expressed in metres, therefore the

newton metre is the unit of torque

Example 1

A spanner of length 0·4 m has a force of 80 N applied to its end as shown in the diagram (Fig. 2.11.3). Calculate the turning moment given by this force.

$$\text{Turning moment} = F \times r = 80 \times 0{\cdot}4 = 32\ \text{N m}$$

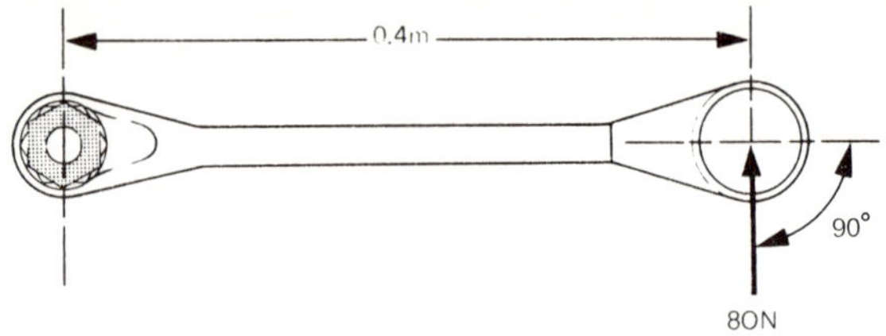

Fig. 2.11.3

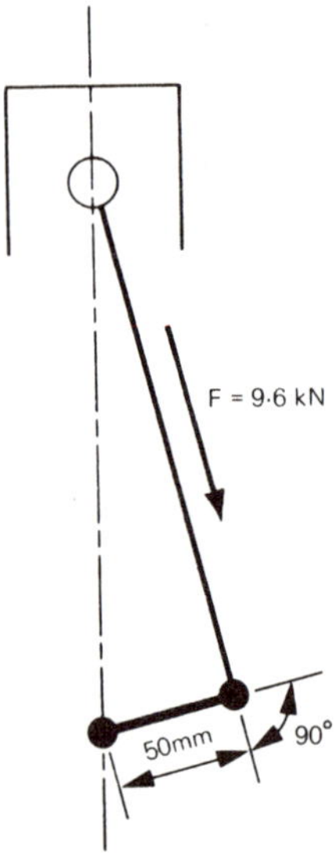

Fig. 2.11.4

Example 2

A connecting rod applies a force of 9·6 kN to a crank of length 50 mm when the crank is positioned to form a right angle to the connecting rod. Calculate the torque at this instant.

Torque = $F \times r$
Force is 9600 N and distance is 0·05 m
Torque = 9600 x 0·05
= 96 x 5
= 480 N m

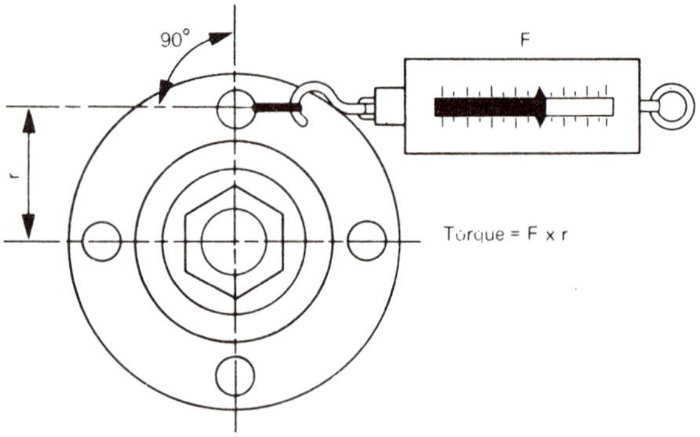

Fig. 2.11.5

Measurement of torque

Occasions occur when the torque applied to a component must be measured. Fig. 2.11.5 shows a spring balance applied to the flange of a final drive pinion to measure the bearing preload. Torque is obtained by multiplying *F* by *r*. Special gauges are available for this job and these are calibrated in N m.

Nuts and bolts of motor vehicles must be tightened the correct amount. Undertightening allows the item to move when it is loaded and can result in the nut working loose, whereas overtightening weakens the bolt and often leads to breakage. The mechanic's experience and skill should enable him to assess the correct tightness, but in certain applications this is unsatisfactory. Assuming the thread is undamaged, the tightness depends on the torque. Where a specific degree of tightness is essential the manufacturer states the torque to which the nut or bolt should be set and this is achieved by a *torque wrench* (Fig. 2.11.6). Various designs are available; the type illustrated is used by setting the wrench to the recommended torque and then tightening the nut until the wrench indicates that the

Torque Wrench

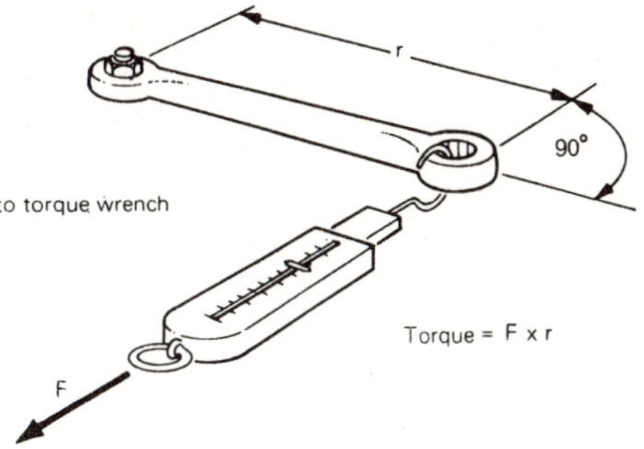

Simple alternative to torque wrench

Fig. 2.11.6

torque has been reached. Typical torque for the tightness of a sparking plug in a cast-iron head is 34 N m.

A spring balance applied to a spanner is an alternative to a torque wrench.

2.12 Principle of moments

The remarkable feat shown in Fig. 2.12.1 illustrates the principle of moments. The car is in equilibrium, i.e. a state of balance, and in this position the sum of the turning moments which tend to rotate the car in a clockwise direction, equals the sum of the moments which tend to rotate the car in the anti-clockwise direction. Phrased in a more precise form:

The principle of moments states that when a body is in equilibrium, then the clockwise moments about any point equal the anti-clockwise moments about the same point.

Taking moments about the jack:

clockwise moments $= F_1 \times a$
anti-clockwise moments $= F_2 \times b$
clockwise moments = anti-clockwise moments
$\therefore \quad F_1 \times a = F_2 \times b$

The jack is taking the whole weight of the car so

$F_1 + F_2$ = total weight of car

In this case it is convenient to use the jack as the point around which the moments are taken, but the principle would apply if any other point was selected.

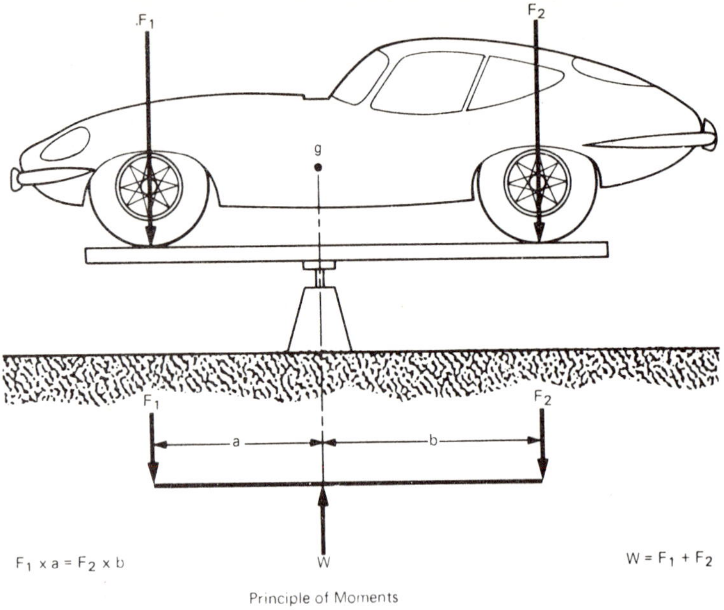

Fig. 2.12.1

Suspending the car in Fig. 2.12.1 by a wire attached to point 'g' would produce a state of balance irrespective of the initial position of the car. Point 'g' is the *'centre of gravity'* and is the theoretical point at which the total mass is considered to act. Instead of each separate mass, i.e. the engine, gearbox, etc., giving a number of individual moments, all the masses can be grouped together to give one mass which is concentrated at the centre of gravity.

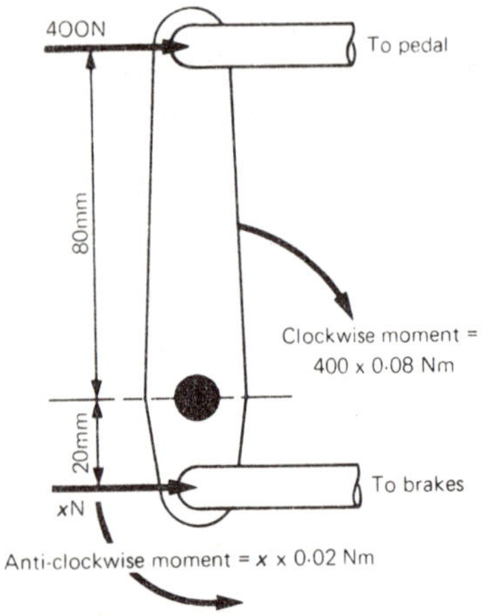

Fig. 2.12.2

Example 1

A brake lever, Fig. 2.12.2, has a force of 400 N applied at a distance of 80 mm from the pivot. What force will act on the rod which connects the lever to the brake?

Take moments about the pivot and let x = required force

Clockwise moment = 400 x 0·08 N m (both moments must be in the same units)

Anti-clockwise moment = x x 0·02 N m

Anti-clockwise moment = clockwise moment

$$x \times 0{\cdot}02 = 400 \times 0{\cdot}08$$

$$x = \frac{400 \times 0{\cdot}08}{0{\cdot}02}$$

$$x = 1600\ \text{N}$$

Example 2

A vehicle having a wheel base of 3 m is balanced on a jack, as shown in Fig. 2.12.3. The jack is positioned 1 m behind the front wheels and applies a force of 9 kN. What will be the force between the tyres and the road when the vehicle is positioned on a level road?

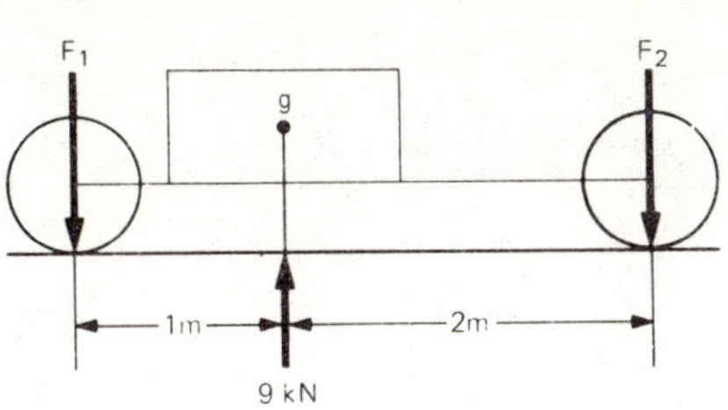

Fig. 2.12.3

Taking moments about the front wheels

Anti-clockwise moment = 9×1
Clockwise moment = $3 \times F_2$
Clockwise moment = Anti-clockwise moment
$3 \times F_2 = 9 \times 1$
$\therefore F_2 = \frac{9}{3} = 3\,\text{kN}$

Since vehicle weighs 9 kN, the force acting on front wheels (F_1) = 9 − 3 = 6 kN.

Example 3

An 'inclined' force acts on a bell crank lever as shown in Fig. 2.12.4. Calculate the force *F*.

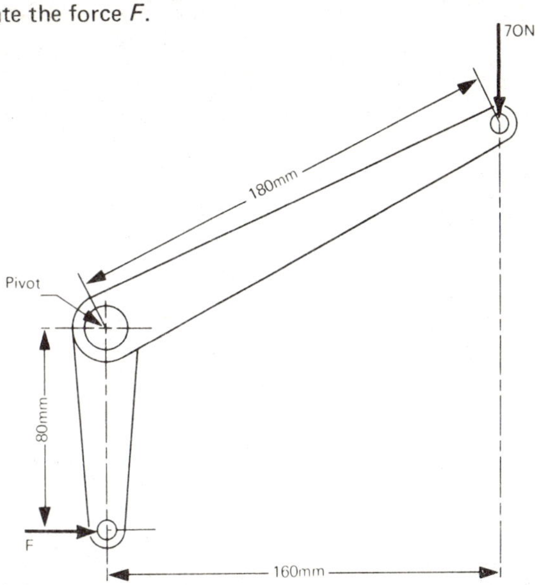

Fig. 2.12.4

Taking moments about the pivot

Clockwise moment = 70 x perpendicular distance
= 70 x 160 N mm
Anti-clockwise moment = *F* x 80 N mm
Anti-clockwise moment = Clockwise moment
$F \times 80 = 70 \times 160$

$$F = \frac{70 \times 160}{80} = 140\ \text{N}$$

Referring back to Fig. 2.11.4 shows how the engine can produce a turning moment. This torque fluctuates during the cycle but with multicylinder engines operating at speed a fairly even torque is obtained. It is this torque which drives the vehicle.

Fig. 2.12.5 shows a driving wheel onto which is applied a torque *T* by means of an axle shaft. Torque is $F \times r$ so a force of *F* is given at the road surface. Grip between the tyre and the road prevents the tyre from skidding and this results in a driving force being applied to the axle casing.

$$\text{Driving force (or tractive effort)} = \frac{\text{Torque (N m)}}{\text{Wheel radius (m)}}\ \text{N.}$$

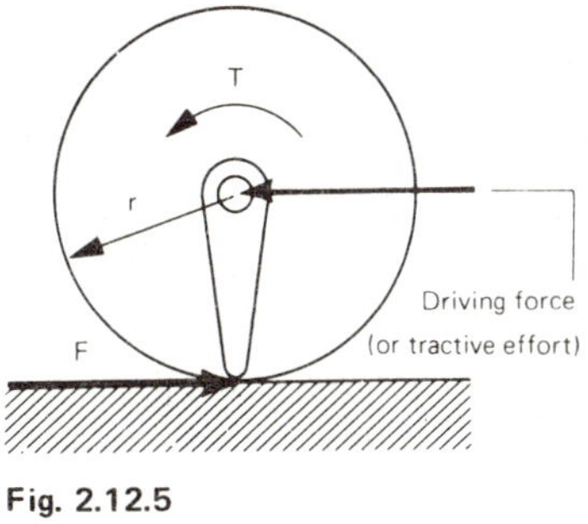

Fig. 2.12.5

Torque at rear wheels

Exercises 2.13

1. The definition:
'that which changes a body's state of rest or of uniform motion in a straight line' applies to
(a) force (c) energy
(b) work (d) power

2. Which one of the following is a unit of force:
newton; kilogramme; tonne; newton metre?
(a) newton (c) joule
(b) kilogramme (d) newton metre

3. The quantity of matter in a body is expressed in:
(a) newtons (c) joules
(b) kilogrammes (d) newton metres

4. The rate of increase in velocity is called
(a) speed (c) movement
(b) acceleration (d) mass

5. What force is required to accelerate a mass of one kilogram at the rate of 1 metre per second per second?
(a) 1 kg (c) 1 N
(b) 9·81 kg (d) 9·81 N

6. The force of the earth's gravity on a body is called its:—
(a) mass (c) weight
(b) acceleration (d) energy

7. Due to the earth's gravity, a body falling freely in a vacuum will accelerate at a rate of about
(a) 1 m/s^2 (c) 32 m/s^2
(b) 9·8 m/s^2 (d) 32·2 m/s^2

8. A jack supports a mass of 0·5 tonne. The force given by the jack is about
(a) 500 kg (c) 51 N
(b) 1000 kg (d) 4·9 kN

9. An object placed on a bench exerts a force of 98·1 N due to gravity. The mass of the object is
(a) 10 N (c) 10 kg
(b) 94 N (d) 98·1 kg

10. What is torque?
(a) Force times distance moved
(b) Acceleration due to gravity
(c) Work applied to a body
(d) A turning moment.

11. A force of 100 N acts as a radius of 0·5 m. The torque is
(a) 0·005 N m (c) 50 N m
(b) 0·05 N m (d) 500 N m

12. A nut has to be tightened to a torque of 80 N m. What force must be applied at a right angle to a spanner of length 0·25 m?

(a) 20 N
(b) 32 N
(c) 200 N
(d) 320 N

13. The theoretical point at which the total mass of a body is considered to act is called the

(a) weight
(b) action point
(c) gravity point
(d) centre of gravity

14. A straight brake lever of length 200 mm has a pivot 50 mm from one end and a force of 84 N is applied at a right angle to the other end.

The force, acting at a right angle, given at the end of the shorter arms is

(a) 26 N
(b) 28 N
(c) 252 N
(d) 336 N

15. When an engine valve is at the midpoint of its lift the force exerted by the valve spring is 330 N. The valve is operated by a pushrod and overhead rocker and in this position the pushrod is parallel to the valve stem and at right angles to the rocker. If the rocker pivot is 30 mm from the axis of the valve stem and 20 mm from the axis of the pushrod, the force in the pushrod is

(a) 220 N
(b) 495 N
(c) 550 N
(d) 825 N

2.14 Energy and work

As stated already:

energy is the capacity for doing work

There are many forms of energy; to name a few – chemical, heat, mechanical and electrical. These are all applicable to motor vehicles and in this book reference is made to these various forms of energy.

Energy cannot be created or destroyed – It can only be converted from one form to another

Internal combustion engines act as a good example to illustrate this statement. Chemical energy enters the engine in the form of fuel; the fuel is burnt to produce heat energy; the heat increases the pressure of the gas and produces mechanical energy. The process does not stop here because the mechanical energy used to drive the car along the road is given back to the atmosphere.

When a force overcomes a resistance and causes movement, *work* is done. Here we are concerned with work in an engineering sense and most probably this differs from the average person's idea of work. Stated in a technical form:

Work = force x distance moved

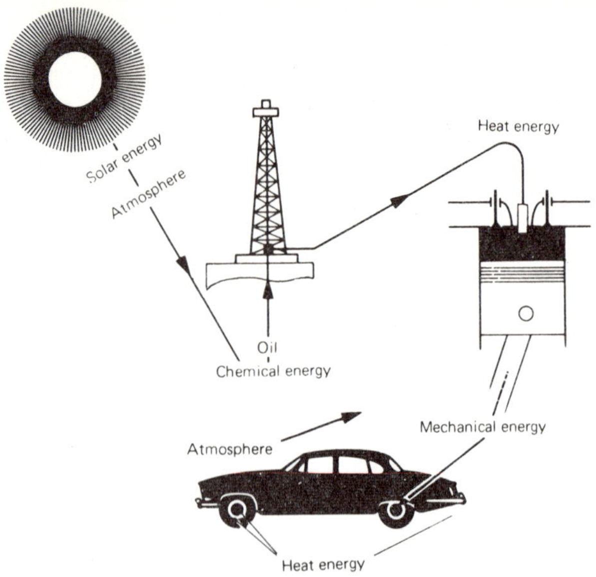

Fig. 2.14.1

Force is measured in newtons; distance is in metres so the product of these two units is a newton metre (N m).

In SI the unit of work is the joule

$$1\text{ J} = 1\text{ N m}$$

(Sometimes it is convenient to state the unit of work as the newton metre – both the newton metre and the joule are acceptable units.)

Work is linked with energy; the unit of energy is the joule.

Fig. 2.14.2 shows a vehicle being driven along a road at a constant speed by a force. The work done is the product of the force and the distance moved.

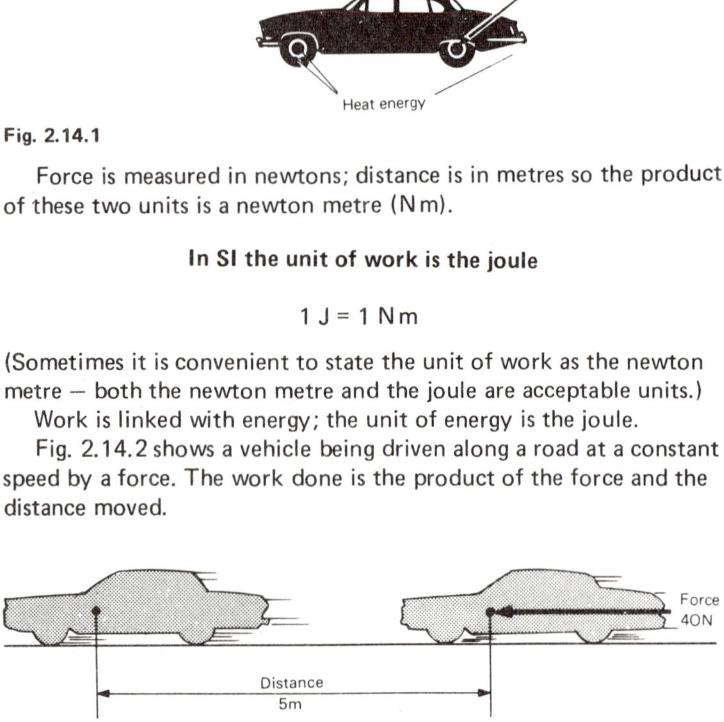

Fig. 2.14.2

Example 1

An engine having a stroke of 70 mm has an average force of 4 kN acting on the piston during the power stroke. Calculate the work done during this stroke.

$$\begin{aligned}\text{Work done} &= \text{Force} \times \text{distance moved}\\ &= 4000 \times 0{\cdot}07\text{ N m}\\ &= 40 \times 7\\ &= 280\text{ N m} = 280\text{ J}\end{aligned}$$

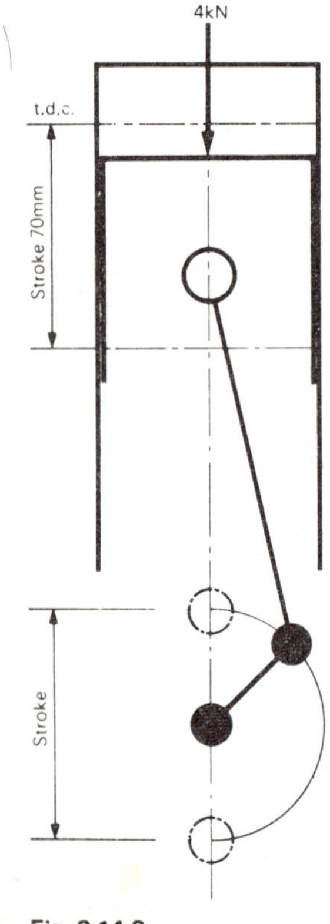

Fig. 2.14.3

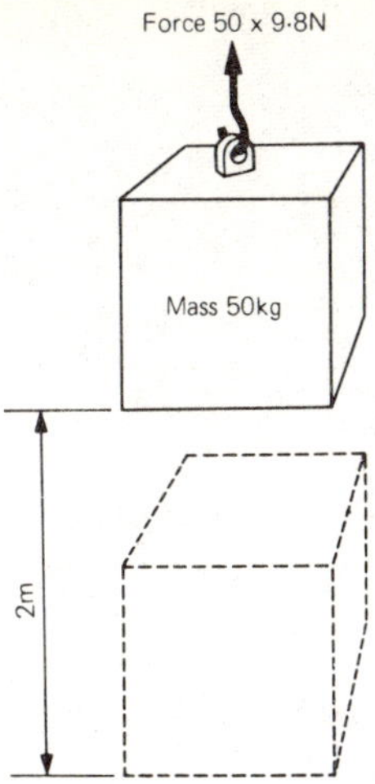

Fig. 2.14.4

Example 2

An engine of mass 50 kg is lifted through a vertical distance of 2 m. Calculate the work done against gravity. Take g as 9·8 m/s^2.

Force required to lift mass = 50 x 9·8
= 5 x 98
= 490 N

Work done = Force x distance moved
= 490 x 2 N m
= 980 J

Example 3

The force given by the brakes when retarding (slowing down) a vehicle is 8 kN. How much energy is converted into heat by the brakes if the retarding force acts over a distance of 30 m?

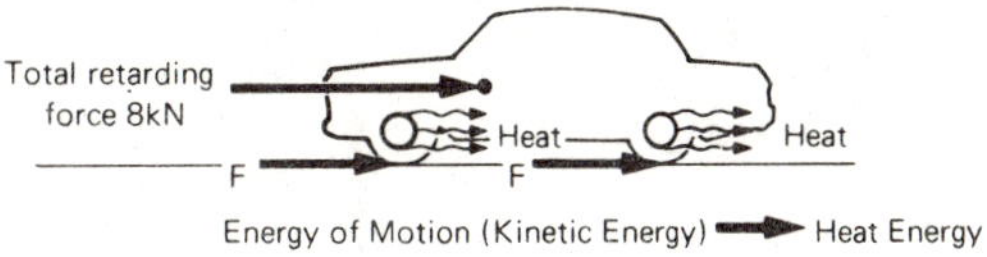

Fig. 2.14.5

Work done by brakes = 8000 x 30
= 240 000 N m
= 240 k N m
= 240 k J

Energy converted by brakes into heat energy = 240 k J

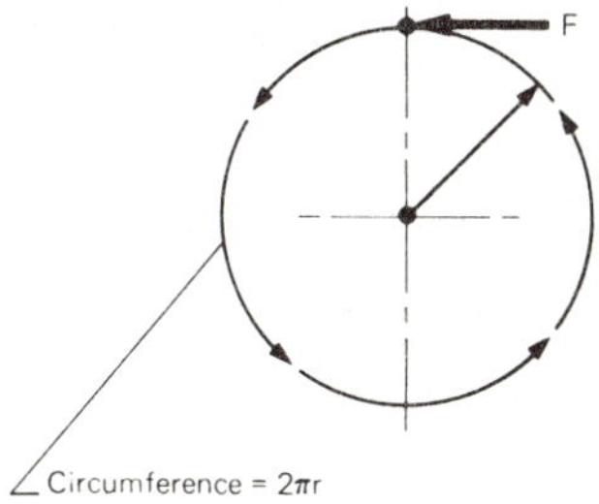

Fig. 2.14.6

Work done by a torque

Work is done when a component is rotated under the action of a torque. A torque of T newton metres may be produced by a force of F newtons acting at a radius of r metres from the centre of rotation. Then, in one revolution, the point of application of the force moves a distance of $2\pi r$.

Work = force x distance moved
= $F \times 2\pi r$ OR $F \times r \times 2\pi$

Since torque is $F \times r$ then

Work done per revolution = $T \times 2\pi$ OR $2\pi T$

Work done in n revolutions = $2\pi n T$

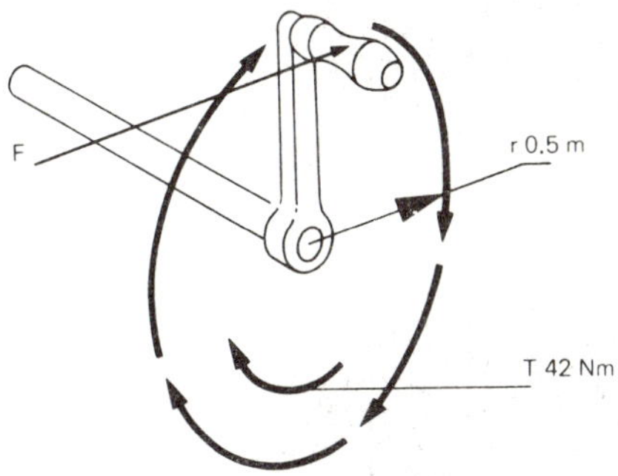

Fig. 2.14.7

Example 4

A torque of 42 N m is applied to a winch handle of radius 0·5 m. Calculate the work done:

(i) per revolution

(ii) per minute when the handle is moved at the rate of 50 rev/min.

(Take π as $\frac{22}{7}$)

(i) Work done per revolution $= 2\pi T$

$$= \frac{2 \times 22 \times \cancel{42}^{\,6}}{\cancel{7}}\ \text{N m}$$

$= 264\ \text{N m} = 264\ \text{J}$

(ii) Work done per minute $= 2\pi\, nT$

$= 264 \times 50\ \text{N m}$

$= 13\,200\ \text{N m} = 13\,200\ \text{J} = 13{\cdot}2\ \text{kJ}$

2.15 Power

The rate at which a machine or engine does work is called its *power*. Work at the rate of one joule per second is the unit of power — the watt.

Power is the rate of doing work and is expressed in watts

The diagram (Fig. 2.15.1) shows an example of the use of the term power. It compares two vehicles: A, with a 'small' engine and B with a 'big' engine. Both vehicles are expected to climb the hill that is shown, but tests show that A will take a longer time — in this case vehicle B is *more powerful* than vehicle A. Assuming the mass of each vehicle is equal, the:

tractive effort of A = tractive effort of B

and work done by A = work done by B

Since B can do this work in half the time then the:

power of B = twice the power of A.

Hence:

$$\text{power} = \frac{\text{force} \times \text{distance moved}}{\text{time}}$$

or $\text{power} = \dfrac{\text{work done}}{\text{time}}$ OR work done in a given time.

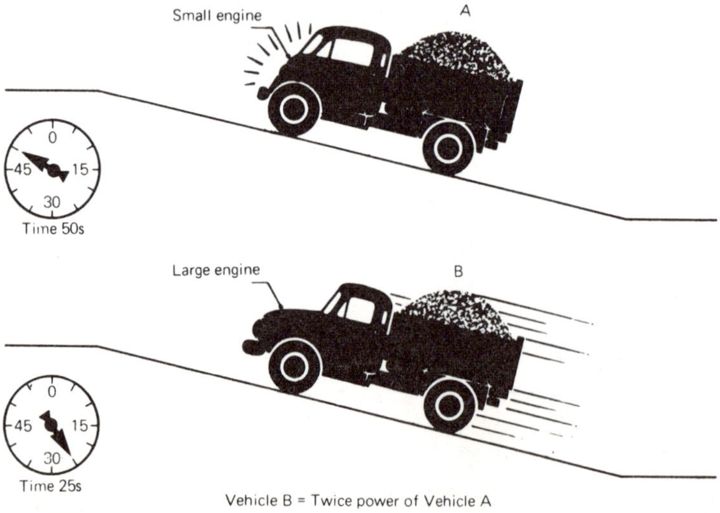

Fig. 2.15.1

Example 1

A vehicle travels a distance of 1·8 km in a time of 2 minutes. If the tractive effort is 200 N, calculate the power used.

Work done = force x distance moved

$= 200 \times 1800$ N m in 2 minutes

$= \dfrac{200 \times 1800}{2 \times 60}$ N m in 1 second

= 3000 N m/s = 3 kN m/s = 3 kJ/s

1 J/s is 1 watt

so power used = 3 kilowatts

which is also written as 3 kW.

Example 2

A tractive effort of 600 N is required to propel a motor car at a speed of 120 km/h. Calculate the power required

$$120 \text{ km/h} = \frac{120\,000}{60 \times 60} = \frac{120\,000}{60 \times 60}$$

$$= \frac{100}{3} \text{ m/s}$$

Work done per second = force x distance moved in 1 second

$$= 600 \times \frac{100}{3} \text{ N m}$$

= 20 000 N m = 20 000 J

Power required = 20 000 W OR 20 kW

Horsepower

The Imperial unit of power is the *horsepower* which is given the symbol hp. This unit was introduced by James Watt to indicate to prospective buyers the power output of his steam engines. In that period the horse provided the power for transport purposes so Watt tested a number of horses and from these results he decided that:

1 horsepower = work at the rate of 33 000 foot pounds per minute

or 1 hp = 33 000 ft lbf/min

For conversion purposes:

1 horsepower ≙ 746 W

2.16 Friction and lubrication

When a block is placed on a horizontal surface and a force is applied to it, the block will not move until the force reaches a value sufficient to overcome an opposing force called *friction* (Fig. 2.16.1). The force required to initially cause sliding is termed the *limiting frictional force* and once this has been overcome a smaller force will keep the block moving.

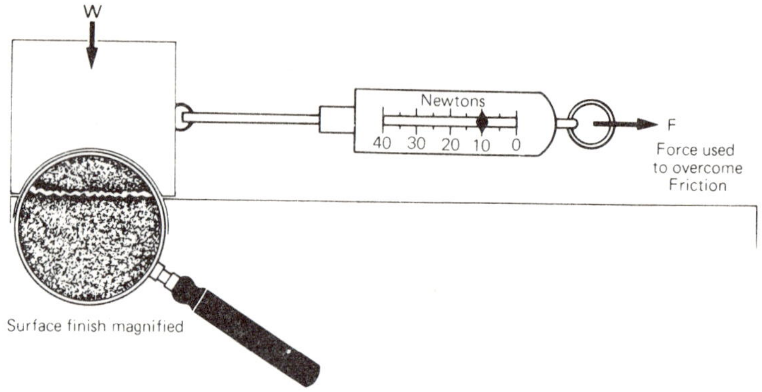

Fig. 2.16.1

Sometimes friction is a disadvantage, but if it did not exist many things that we do would be impossible, e.g. walking — no grip between the shoe and the ground would mean that a new method of propulsion would be needed.

With motor vehicles it is possible to consider the effects of friction and group these effects into situations where friction is (a) an advantage and (b) a disadvantage.

(a) *Situations where friction is usefully employed*

(i) *clutch* — drive is made possible by the friction between the plates
(ii) *brakes* — energy of motion is converted to heat by rubbing one surface against another
(iii) *tyres* — the wheel rim drives the tyre by friction. The tyre 'grips' the road.
(iv) *fan belt* — drive to the fan and generator is made possible by friction.
(v) *steering wheel* — friction between the driver's hand and the steering wheel.

(b) *Situations where friction is disadvantageous*

(i) *piston* — for high efficiency the piston should slide easily in the cylinder
(ii) *bearings* — shafts must turn as freely as possible — friction in bearings results in a reduction in the power applied to drive the vehicle. 'Rolling friction' is much smaller than sliding friction, so ball or roller bearings are efficient types of bearing.

In cases where friction is essential the designer selects his material to give the appropriate grip between the surfaces. All the time this friction is maintained the component will perform its task, but if friction is decreased due to wear or other reasons then slip will occur.

The frictional resistance between two surfaces depends on the:

(a) *nature of the materials* – there is more friction between asbestos and iron than between steel and bronze.

(b) *condition of the surfaces* – smooth or rough; wet or dry.

(c) *force pressing the surfaces together* – friction is proportional to the force, i.e. when the force thrusting one surface against another is doubled, the friction is doubled.

N.B. Under normal conditions the area in contact does not affect the friction.

Lubrication

It is believed that friction is caused by the interlocking action of the two surfaces (fig. 2.15.1) and also by the attraction of one material to another. Placing two surfaces together causes the load to be supported by small 'crests'. If one surface is now slid over the other, the crests will attain a very high temperature and will weld the two materials together. Further movement will break these welds, 'tear' the surface, and cause rapid wear. A reduction in the energy lost to friction and longer component life can be obtained if the surfaces can be separated; this is achieved by lubrication.

A lubricant introduced between the surfaces may be considered as a series of liquid globular particles which easily slide over each other. Provided there is no metal to metal contact, any friction which exists will be caused by the resistance of one particle to leave its neighbour – this resistance is called the *viscosity* of a lubricant.

The type of lubricant is governed by the conditions under which it operates and some of the duties of a lubricant can be appreciated when the following are considered.

Engine. The oil must resist any change of property caused by heat or corrosive acids. It must also flow easily to act as a coolant.

Final drive. A very high gear tooth loading prevents an oil film being maintained the whole time, so the oil must provide an alternative method of lubrication during the period that the oil film is broken.

These properties can be introduced or improved by inserting into the oil special chemicals called *'additives'.* Since each group of additives is intended to meet a special demand, then it is essential that the correct oil is used for each vehicle component.

Exercises 2.17

1. What is 'energy'?
(a) The capacity for doing work
(b) Force times perpendicular distance
(c) The rate of doing work
(d) Force times distance moved

2. The SI unit of energy is the
(a) newton
(b) joule
(c) watt
(d) ft lbf

3. What is the technical meaning of the term 'work'?
(a) Force times distance moved
(b) Distance moved in a given time
(c) Force times radius
(d) Force obtained from energy

4. The SI unit of work is the
(a) joule
(b) watt
(c) kilogramme
(d) ft lbf

5. An engine converts energy. In which order does the energy change to complete the process between the time that fuel is supplied, and power is produced?
(a) Heat energy, mechanical energy, chemical energy
(b) Heat energy, chemical energy, mechanical energy
(c) Chemical energy, mechanical energy, heat energy
(d) Chemical energy, heat energy, mechanical energy

6. How much work is done when a force of 200 N is required to lift an object a distance of 5 m?
(a) 1kJ
(b) 40 N m
(c) 40 J
(d) 1000 mN

7. An engine having a mass of 200 kg is lifted vertically a distance of 2 m. What work does this represent? (Take g as 10 m/s^2.)
(a) 40 J
(b) 400 J
(c) 1 kJ
(d) 4 kJ

8. The work done in moving a crate along the ground for a distance of 5 m is 4 kJ. How much heat energy is produced?
(a) 20 J
(b) 800 J
(c) 4 kJ
(d) 20 kJ

9. A torque of 14 N m is applied to a steering wheel of radius 0·4 m. The work done when the steering wheel is moved through two revolutions is
(a) 35 J
(b) 56 J
(c) 176 J
(d) 1·232 kJ

10. What is meant by the term 'power'?
(a) The rate of doing work
(b) The capacity for doing work
(c) Force times perpendicular distance
(d) Force times distance moved

11. The SI unit of power is the
(a) watt
(b) joule
(c) newton
(d) horsepower

12. A force of 200 N acts over a distance of 5 m for a time of 4 s. This represents a power of
(a) 10 W
(b) 250 W
(b) 1000 W
(d) 4000 W

13. A force of 180 N acts over a distance of 2 km for a time of 2 minutes. This represents a power of
(a) 3 W (c) 3 kW
(b) 180 W (d) 180 kW

14. A vehicle of weight 9 kN is raised through a vertical distance of 2 m in 30 s. This represents a power of
(a) 0·6 W (c) 600 W
(b) 150 W (d) 1·08 MW

15. A vehicle of mass 4500 kg is raised through a vertical distance of 2 m in 1·5 minutes. If g is taken as 10 m/s^2, this represents a power of
(a) 10 W (c) 1 kW
(b) 100 W (d) 60 kW

16. What is friction?
(a) A form of heat energy
(b) Work stored in a body
(c) A force which opposes motion
(d) The change of state of a substance

17. In which one of the following applications is friction regarded as a disadvantage
(a) Piston (c) Brake
(b) Clutch (d) Fan belt

18. Assuming that the operating conditions are normal, which one of the following factors does **not** affect the frictional resistance?
(a) Nature of the surfaces
(b) Area of surfaces
(c) Condition of surfaces
(d) Force pressing surfaces together

19. Which words are required to make the following statement correct: high friction is obtained when the surface is . . . . and . . . ?
(a) smooth . . . . dry (c) rough . . . . dry
(b) smooth . . . . wet (d) rough . . . . wet

20. Why does friction cause wear? Because the
(a) heat causes the surfaces to harden
(b) heat expels the air from the surfaces
(c) surfaces cannot obtain a supply of air
(d) surfaces interlock, heat, weld and tear

21. How does a normal lubricant reduce friction?
The lubricant
(a) allows air to pass between the surfaces
(b) keeps the surfaces apart
(c) smooths the surface of the metal
(d) prevents the generation of heat

2.18 Machines

A *machine* is a device for transmitting energy – in most cases it enables a large force to be given at the 'output' end of the machine when only a small force is applied at the 'input'. Some machines are extremely simple; and examples of these are the humble spanner, the hammer and various kinds of levers. Some are exceedingly complex, like some of the automatic machine tools used for making motor vehicle components. We consider here the basic principles of some of the simple types of machine.

The lever

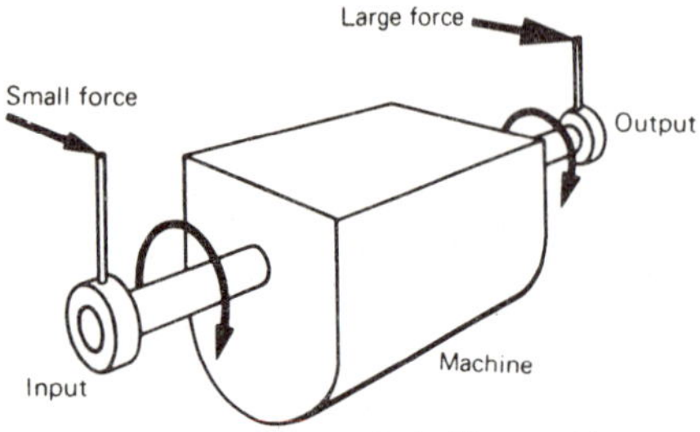

Multiplication of torque obtained from machine

Fig. 2.18.1

The lever is essentially a bar, either straight or cranked (bent) which is pivoted at some point called the *fulcrum.* An *effort* applied at some point on the lever overcomes a *load* at some point on the lever. There are three basic types or 'orders' of lever and Fig. 2.18.2 shows these

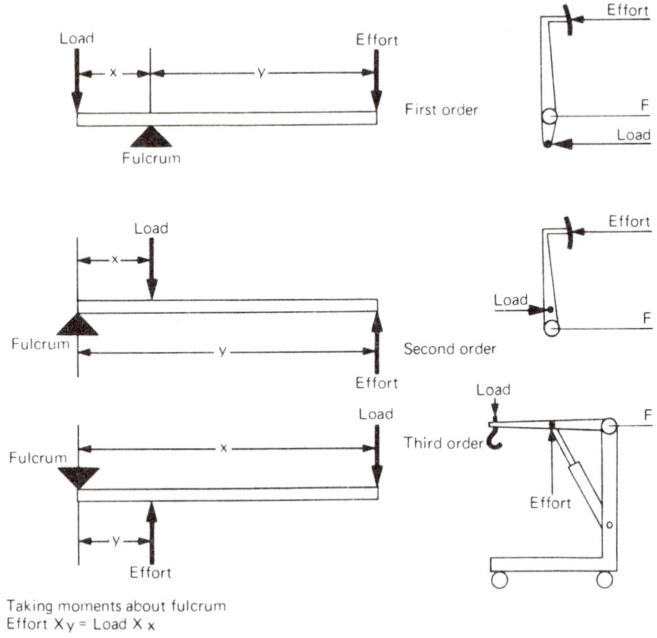

The lever

Fig. 2.18.2

Referring to Fig. 2.18.2 this shows that

$$\text{LOAD} \times x = \text{EFFORT} \times y$$

$$\therefore \frac{\text{LOAD}}{\text{EFFORT}} = \frac{y}{x}$$

The ratio $\frac{\text{LOAD}}{\text{EFFORT}}$ is called the *force ratio* or *mechanical advantage.*

To obtain a large force ratio with a lever, the distance 'y' must be much longer than the distance '*x*'. At first it appears that we are getting 'something for nothing' but this idea is modified when we consider the movement of the effort in respect to the movement of the load. This relationship is called the *movement ratio* or *velocity ratio.* To summarise:

$$\textbf{Force ratio} = \frac{\textbf{load}}{\textbf{effort}}$$

$$\textbf{Movement ratio} = \frac{\textbf{movement of effort}}{\textbf{movement of load}}$$

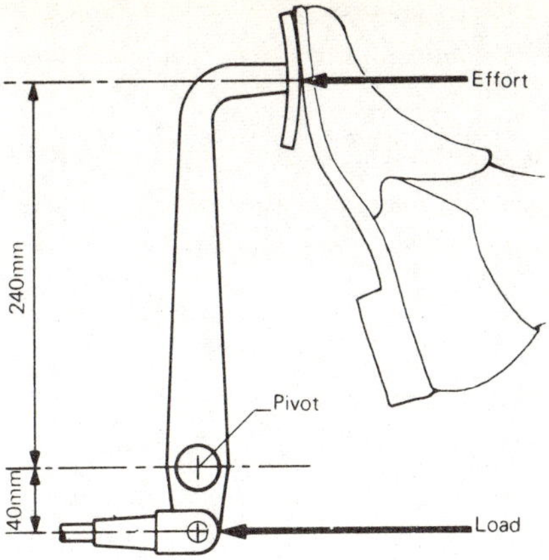

Fig. 2.18.3

If friction was not present in the machine then:

Force ratio = movement ratio

Example 1

Fig. 2.18.3 shows a lever which is acting as a brake pedal. Neglecting friction calculate:
(a) the effort required to produce a force of 1·2 kN on the brake rod
(b) the movement of the rod when the pedal pad moves 84 mm.

(a) Taking moments about the pivot

$$\text{effort} \times 240 = 1{\cdot}2 \times 40$$
$$\therefore \text{ effort} = \frac{1{\cdot}2 \times 40}{240} = 0{\cdot}2 \text{ kN}$$

(The ratio $\dfrac{\text{distance between effort and fulcrum}}{\text{distance between load and fulcrum}}$ is sometimes called the leverage. Effort x leverage = load)

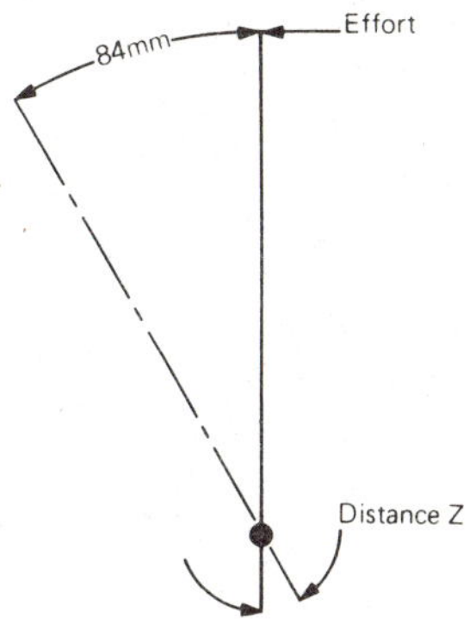

Fig. 2.18.4

(b) Distance z is proportional to the distance from the pivot.
Effort applied at point 240 mm from pivot
Load applied at point 40 mm from pivot

$$\text{movement ratio} = 6$$
$$\text{movement ratio} = \frac{\text{movement of effort}}{\text{movement of load}}$$
$$6 = \frac{84}{\text{movement of load}}$$
$$\therefore \text{ movement of load} = \frac{84}{6} = 14 \text{ mm}$$

In this example it will be seen that:

work input = force x distance moved
= 16·8 kNmm = 16·8 Nm
work output = 1·2 x 14
= 16·8 kNmm = 16·8 Nm

If friction is neglected:

work input = work output

This is an assumption which is never justified in practice since friction always opposes movement.

Efficiency

The effect of friction means that some of the work put into a machine is converted into heat and lost. So:

work output is always less than the work input

To indicate the extent of these internal friction losses the term *efficiency* is used. This can be applied to give the expressions:

$$\textbf{efficiency} = \frac{\textbf{work output}}{\textbf{work input}}$$

or

$$\text{efficiency} = \frac{\text{load x movement of load}}{\text{effort x movement of effort}}$$

$$= \frac{\text{load}}{\text{effort}} \div \frac{\text{movement of effort}}{\text{movement of load}}$$

$$= \frac{\dfrac{\text{load}}{\text{effort}}}{\dfrac{\text{movement of effort}}{\text{movement of load}}}$$

$$\textbf{efficiency} = \frac{\textbf{force ratio}}{\textbf{movement ratio}} \quad \text{OR} \quad \frac{\text{mechanical advantage}}{\text{velocity ratio}}$$

Expressed in this way, efficiency will have a value less than one. To express efficiency as a percentage, multiply above values by 100.

The screw jack

A lifting jack is normally a part of a vehicle's tool kit and Fig. 2.18.5 shows one type; a screw jack. Effort applied by a spanner rotates the screw in the nut. Attached to the nut is an arm which engages in a socket welded to the frame of the car. Besides providing the lifting action, the arm prevents rotation of the nut. During a single revolution of the screw the nut will be moved a distance equal to the *lead* of the screw. With a single start thread the *lead* is equal to the *pitch*.

In one revolution of the screw:

effort moves $2\pi r$

$$\text{Since movement ratio} = \frac{\text{movement of effort}}{\text{movement of load}}$$

$$= \frac{2\pi r}{\text{lead of screw}}$$

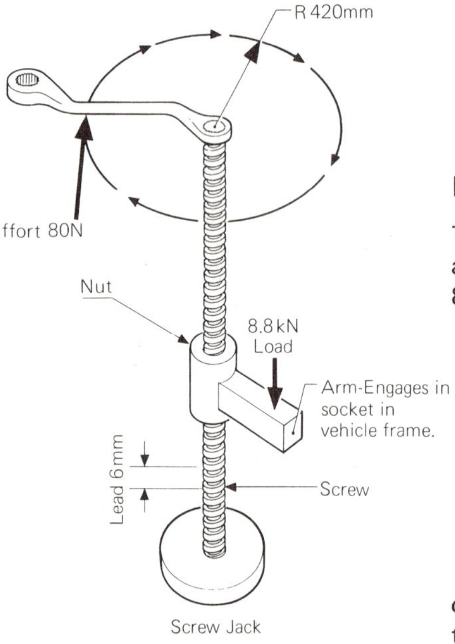

Fig. 2.18.5

Example 2

The screw of a lifting jack has a lead of 6 mm and an effort of 80 N applied at a radius of 420 mm is just sufficient to move a load of 8·8 kN. Calculate the efficiency of the jack.

$$\text{Force ratio} = \frac{\text{load}}{\text{effort}} = \frac{8800}{80} = 110$$

$$\text{Movement ratio} = \frac{2\pi r}{\text{lead}} = \frac{2 \times 22 \times 420}{7 \times 6} = 440$$

$$\text{efficiency} = \frac{\text{force ratio}}{\text{movement ratio}} = \frac{110}{440} = 0{\cdot}25 \text{ or } 25\%$$

This low value means that 75% of the work input is used to overcome friction in the machine. When the efficiency is less than 50% the machine is irreversible and the load is self-sustaining, i.e. the load on the nut will not rotate the screw.

Gearing

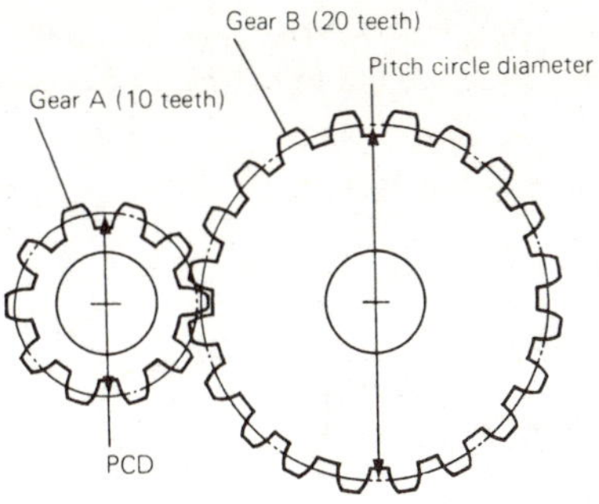

Fig. 2.18.6

Torque could be transmitted by pressing two smooth wheels together, but friction between the wheels would be insufficient to prevent slip when a high driving torque is applied. To obtain a positive drive, teeth are cut on the surface of the wheels to form gears. The teeth are formed above and below the curved surface of the smooth wheels so that the gears would have the same movement ratio as the wheels. The diameter of the wheel, which is equivalent to the diameter of the gear on this basis, is called the *pitch circle diameter* (p.c.d.).

The comparative diameters of two meshing gears governs the relative speeds, e.g. if the p.c.d. of gear B in Fig. 2.18.6 is double that of gear A then the smaller gear will rotate twice as fast as gear B. This speed relationship is similar to movement ratio – it indicates the movement of the input relative to the output and is generally called the *gear ratio*.

$$\text{Gear ratio} = \frac{\text{revolutions of driver}}{\text{revolutions of driven}}$$

$$= \frac{\text{p.c.d. of driven}}{\text{p.c.d. of driver}}$$

$$= \frac{\text{number of teeth on driven}}{\text{number of teeth on driver}} \text{ OR } \frac{\text{DRIVEN}}{\text{DRIVER}}$$

Example

A pinion with 26 teeth drives a gear wheel having 78 teeth. Calculate the speed of the output shaft when the input turns at 2700 rev/min.

$$\text{Gear ratio} = \frac{\text{driven}}{\text{driver}}$$

$$= \frac{78}{26} = 3$$

The ratio is stated as '3 to 1', meaning that three turns of the input are required to rotate the output shaft once.

Hence:

$$\text{speed of output} = \frac{\text{speed of input}}{\text{gear ratio}} = \frac{2700}{3} = 900 \text{ rev/min.}$$

Example

An output shaft of a gearbox consisting of a pair of gears turns 12 revolutions for every 9 revolutions made by the input shaft. If the p.c.d. of the driven gear is 75 mm calculate the p.c.d. of the other gear.

$$\text{Ratio} = \frac{\text{revolutions of driver}}{\text{revolutions of driven}} = \frac{9}{12} = 0{\cdot}75:1$$

To drive output faster the driving gear must be larger

$$\text{Ratio} = \frac{\text{p.c.d. of driven}}{\text{p.c.d. of driver}}$$

$$\text{p.c.d. of driver} = \frac{\text{p.c.d. of driven}}{\text{ratio}} = \frac{75}{0{\cdot}75} = 100 \text{ mm}$$

A gear layout which causes the output shaft to rotate slower than the input would be called a *reduction gear.* Similarly an increase of speed is termed an *overdrive.*

Torque ratio

As applied to gearing the expression 'force ratio' is referred to as the '*torque ratio*'.

$$\text{Torque ratio} = \frac{\text{torque on output shaft}}{\text{torque on input shaft}}$$

When the efficiency is 100%, the:

torque ratio = speed ratio = gear ratio

Example

A gearbox provides a reduction of 4:1. What torque acts on the output shaft when the input torque is 70 N m? (Neglect frictional losses.)

Gear ratio = torque ratio = 4:1

Since the speed is decreased, the torque is increased.

$$\begin{aligned} \therefore \text{ Torque on output shaft} &= \text{input torque x ratio} \\ &= 70 \times 4 \\ &= 280 \text{ N m} \end{aligned}$$

Example

A gearbox has a ratio of 5:1, an output torque of 70 Nm and an input speed of 3000 rev/min. Assuming an efficiency of 100% calculate the

(a) output speed
(b) output torque
(c) work input per minute
(d) work output per minute
(e) power input and output

(a) $\text{Output speed} = \frac{\text{input speed}}{\text{ratio}} = \frac{3000}{5} = 600 \text{ rev/min}$

(b) Output torque = input torque x ratio = 70 x 5 = 350 N m

(c) Work input per minute = $2\pi nT$. Where n = speed rev/min
T = torque N m

$$\begin{aligned} &= \frac{2 \times 22 \times 3000 \times \overset{10}{\cancel{70}}}{\cancel{7}} \text{ N m} \\ &= 1\,320\,000 \text{ N m} \\ &= 1{\cdot}32 \text{ MJ} \end{aligned}$$

(d) Work output per minute = $2\pi nT$

$$= \frac{2 \times 22 \times 600 \times 350}{7}$$

$= 1\ 320\ 000$ N m
$= 1{\cdot}32$ MJ

(e) Power input $= 1{\cdot}32$ MW
Power output $= 1{\cdot}32$ MW

When the machine has an efficiency of 100% the power does not change. Examples show that the increase in torque is achieved by a sacrifice in speed, i.e.

Gearing efficiency

input speed x input torque = output speed x output torque when efficiency is 100%

Previous problems on gearing have been based on the assumption that the gearbox is an ideal machine, i.e. a machine in which friction is not present. Such an arrangement does not exist in practice so we must accept that the work output is always less than the work input.

As in the case of the screw jack and lever, friction does not alter the movement ratio, it only affects the force ratio or torque ratio.

Example

A gearbox providing a ratio of 8:1 has an efficiency of 80% when a torque of 200 N m is applied to an input shaft rotating at 4000 rev/min. Calculate the output shaft

(a) speed
(b) torque

(a) $$\text{Speed of output} = \frac{\text{Input speed}}{\text{ratio}} = \frac{4000}{8} = 500 \text{ rev/min}$$

(b) When efficiency is 100%

$$\text{output shaft torque} = \text{input torque} \times \text{ratio}$$
$$= 200 \times 8$$
$$= 1600 \text{ N m}$$

When efficiency is 80%, output torque is less

$$\text{output shaft torque} = 1600 \times \frac{80}{100}$$
$$= 1600 \times 0{\cdot}8 = 1280 \text{ N m}$$

Hydraulic machines

A vehicle braking system and various items of garage equipment use the principle of a hydraulic machine. Low friction losses make the system very efficient and when this is coupled with advantages gained by using flexible pipes instead of rigid linkages, then the appeal of hydraulic equipment can be appreciated.

Fig. 2.18.7 shows the layout of a simple hydraulic lift. A small plunger pump displaces fluid to a large ram cylinder which in this case operates a vehicle lift. As with any machine, to obtain a large force-ratio the movement of the effort is much greater than the movement of the load.

Liquid is incompressible, so operation of the plunger will move the liquid from the small cylinder to the ram cylinder. Since the

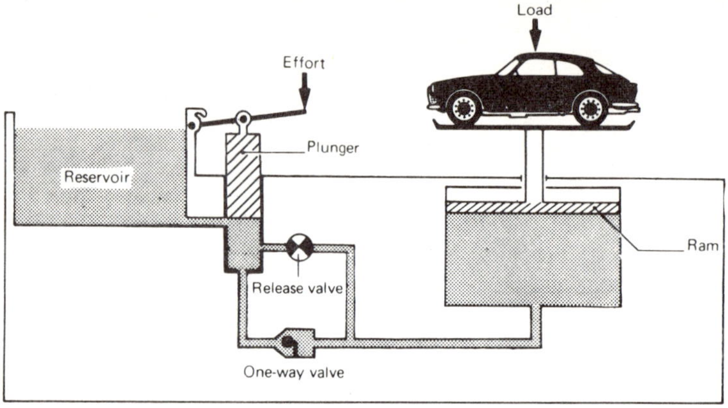

Fig. 2.18.7

load is acting against the ram piston, the liquid has a tendency to compress: the force acting on a fluid in this manner is called *pressure.* The area on which the force acts governs the pressure, so the definition used is:

Pressure is the force acting on a unit of area

The force is expressed in newtons, the area in square metres, so

pressure is measured in N/m^2

Other units of pressure will be experienced by the reader:

pascal (symbol Pa) is equal to 1 N/m^2
bar (symbol bar) is equal to 100 000 N/m^2 (10^5 N/m^2)

The Imperial system used 'point per square inch' as the unit of pressure:

$$1 \text{ lbf/in}^2 \simeq 7 \text{ kN/m}^2$$

To avoid confusion we shall use the basic derived unit; the newton per square metre.

Now returning to Fig. 2.18.7, consider the plunger produces a pressure in the operating cylinder of 700 kN/m^2 (7 bar). When friction is neglected, this pressure will act throughout the system so if the ram has an area of:

0·5 m^2 the thrust on the ram is 350 kN
1·0 m^2 the thrust on the ram is 700 kN

From this we may conclude that a large force-ratio is obtained when the *area* of the ram is much greater than the *area* of the plunger. Hence

$$\textbf{Force ratio} = \frac{\textbf{area of ram}}{\textbf{area of plunger}}$$

Example

A hydraulic press has a ram area which is 1000 times greater than the plunger area. Neglecting friction, what force will act on the ram when a force of 800 N is applied to the plunger?

$$\text{Force applied at ram} = 800 \times 1000 \text{ N} = 800\,000 = 800 \text{ kN}$$

Hydraulic braking system

Fig. 2.18.8 shows the layout of a simple braking system. Force applied by the piston in the master cylinder is transmitted through the brake lines to the wheel cylinders. Pressure is the same though-out the system so the brakes will start to apply on each wheel at about the same time, even if the movement of one wheel cylinder is greater than the others.

Leakage due to fracture of a pipe will prevent the build-up of pressure and this will result in failure of all brakes unless special provision is made.

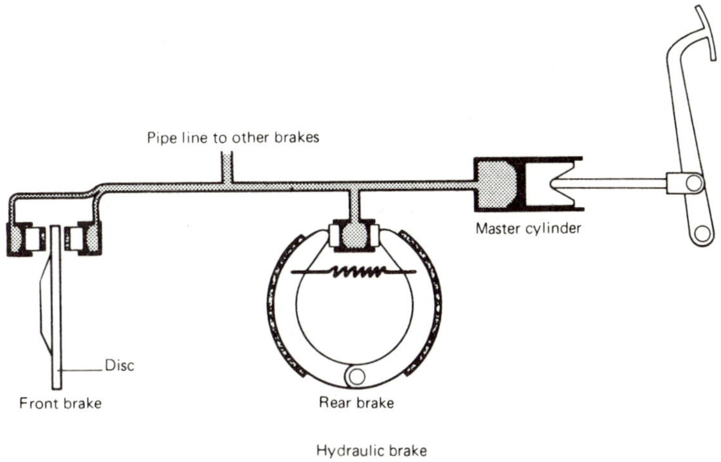

Fig. 2.18.8

Exercises 2.19

1. During the lubrication process the oil particles move over each other. The resistance of one particle to leave adjacent particles of the lubricant is called
(a) oiliness (c) flow
(b) additive (d) viscosity

2. What is a machine? A device for
(a) transmitting energy (c) increasing the work done
(b) generating energy (d) reducing load and effort

3. Another name for a pivot is
(a) effort (c) centre of gravity
(b) fulcrum (d) turning moment

4. Which one of the following expressions gives the force ratio?
(a) effort + load
(b) effort x load
(c) $\frac{\text{effort}}{\text{load}}$
(d) $\frac{\text{load}}{\text{effort}}$

5. Which one of the following expressions gives the movement ratio?
(a) movement of effort + movement of load
(b) movement of effort x movement of load
(c) $\frac{\text{movement of effort}}{\text{movement of load}}$
(d) $\frac{\text{movement of load}}{\text{movement of effort}}$

6. If friction is not present in a machine then 'force ratio' equals
(a) movement ratio
(b) $\frac{1}{\text{movement ratio}}$
(c) efficiency + movement ratio
(d) movement ratio x 100

7. Which words are required to make the following statement correct?
In an actual machine the . . . . is always greater than the . . . .
(a) effort . . . . load
(b) load . . . . effort
(c) force ratio . . . . movement ratio
(d) movement ratio . . . . force ratio

8. A straight lever of length 500 mm is pivoted at a point 50 mm from the end at which the load is applied. Neglecting friction what load will be given if an effort of 90 N is applied at a right angle to the end of the long arm of the lever?
(a) 9 N
(b) 10 N
(c) 810 N
(d) 900 N

9. A crowbar gives a leverage of 30:1. If the efficiency is 90%, an effort of 90 N will lift a load of
(a) 2·7 N
(b) 3·3 N
(c) 2·43 kN
(d) 3 kN

10. A machine has a movement ratio of 8 and a force ratio of 6. The efficiency is
(a) 1·3% (b) 13·3% (c) 48% (d) 75%

11. A machine has a force ratio of 20 and a movement ratio of 40. The efficiency is
(a) 0·5 (b) 2·0 (c) 20 (d) 60

12. A crowbar gives a leverage of 30:1. If the efficiency is 80%, a load of 3 kN will require an effort of
(a) 80 N (b) 125 N (c) 72 kN (d) 112·5 kN

13. A machine has a work input of 600 kJ and a work output of 300 kJ. The efficiency is
(a) 2% (b) 50% (c) 90% (d) 300%

14. The screw of a simple lifting jack has a lead of 10 mm. If the spanner operating the screw has an effective radius of 0·21 m, the movement ratio is
(a) 0·021 (b) 0·132 (c) 2·1 (d) 132

15. If r is the effective length of a spanner which operates a simple screw jack, then one revolution of the spanner will cause the effort to move
(a) $2\pi r$ (b) πr^2 (c) $2\pi r^2$
(d) a distance equal to the lead of the screw

16. Moving a spanner which operates a simple screw jack through one revolution will cause the load to move
(a) $2\pi r$ (b) πr^2 (c) $2\pi r^2$
(d) a distance equal to the lead of the screw

17. What would be the effect if the efficiency of a vehicle jack was greater than 50%?
The jack would
(a) require a large effort to lift a small load
(b) have a force ratio greater than the movement ratio
(c) not allow the vehicle to be lowered
(d) not sustain the load

18. A pinion with 25 teeth drives a gearwheel having 75 teeth. When the input speed and torque is 2100 rev/min and 60 N m respectively, the output speed and torque is
(a) 700 rev/min and 180 N m
(b) 6300 rev/min and 20 N m
(c) 700 rev/min and 180 N m if efficiency is 100%
(d) 6300 rev/min and 20 N m if efficiency is 100%

19. A gearbox has an efficiency of 90% when the input speed is 4320 rev/min and output speed is 1200 rev/min. The gear ratio is
(a) 1:3·6 (c) 3·6:1
(b) 3·24:1 (d) 4:1

20. A pinion of a pair of gears has 40 teeth and rotates at 700 rev/min. If the output shaft turns at 280 rev/min, the number of teeth on the meshing gear is
(a) 10 (b) 16 (c) 100 (d) 160

21. An output shaft of a gearbox consisting of a pair of gears turns 6 revolutions for every 15 revolutions made by the input shaft. If the p.c.d. of the driving gear is 90 mm, the p.c.d. of the mating gear is
(a) 22·5 mm (c) 225 mm
(b) 36 mm (d) 360 mm

22. The torque at the output shaft of a gearbox is 120 N m when the input torque is 50 N m. If the gear ratio is 3:1 the gearbox efficiency is
(a) 41·6% (b) 58·3% (c) 71·4% (d) 80%

23. A vehicle is fitted with a simple hydraulically operated brake system. If the left-hand front hose fractures, the effect will be that the
(a) vehicle will pull to the left
(b) vehicle will pull to the right
(c) rear brakes only will operate
(d) brakes will not operate on any wheel.

24. One advantage of a simple hydraulic brake is that the
(a) system does not require any maintenance
(b) pressure is the same throughout the system
(c) master cylinder moves a large distance
(d) front brakes apply before the rear.

2 SCIENCE (c) Materials

A wide range of materials is used in modern motor vehicles and while all students should have some knowledge of the materials used and their properties it will be possible to give only a very brief introduction to the subject here.

2.20 Properties of materials

Strength

The ability of a material to withstand loads without breaking. It is measured in terms of the *ultimate stress.* Stress is the resistance of a material to deform when it is subjected to a load.

$$\text{Stress} = \frac{\text{load (N)}}{\text{cross sectional area (m}^2\text{)}}\ \text{N/m}^2$$

It will be seen that the unit is the same as that used for pressure. Other units are sometimes used: they are:—

$$\text{bar} = 100\,000\ \text{N/m}^2 \quad \text{OR} \quad (10^5\ \text{N/m}^2)$$
$$\text{pascal} = 1\ \text{N/m}^2$$

The Imperial units of stress are the 'lbf/in^2' and 'tonf/in^2'. Fig. 2.20.1 shows the main types of stress.

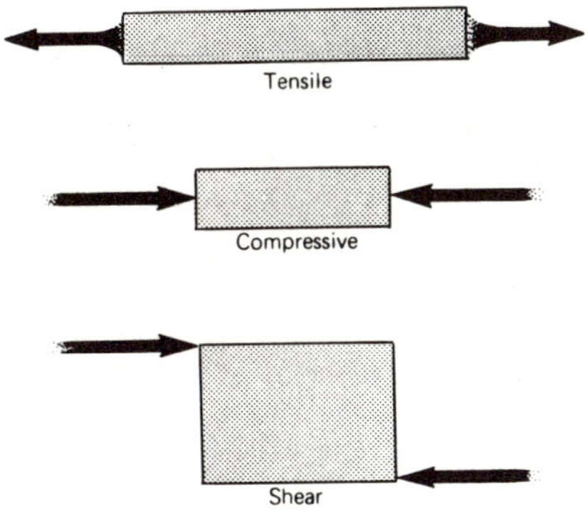

Fig. 2.20.1

Elasticity

When loads are applied to a component there is some deformation or change of shape which is called *strain.* If the strain disappears on removing the load the material is said to be *elastic*, i.e. *elasticity* is the property by which a material is able to regain its original size and shape after being deformed by applied loads.

Most materials are elastic below a certain intensity of stress which is known as the *elastic limit.* If the stress exceeds this limit there will be some *permanent strain* on removing the load.

Plasticity

The property of a material by which it retains any deformation produced by loads after the loads are removed. It is the reverse of elasticity.

Ductility

A material which is plastic under tensile loads is said to be *ductile.* It can undergo considerable permanent stretch *without breaking.*

Ductility is particularly important in manufacturing processes involving bending and pressing and in the manufacture of wires.

Malleability

A material which is plastic under compressive loads is said to be *malleable.* Thus malleability is the property by which a material can undergo considerable permanent deformation *without breaking* when hammered or rolled.

Brittleness

Some materials break with little or no deformation. These are said to be *brittle*, and are particularly liable to break under suddenly applied (or 'shock') loads.

Toughness

The resistance of a material to fracture, measured by the amount of energy required to break a standard specimen. It is applied particularly to the ability to resist breaking under shock loads.

Hardness

The resistance of a material to penetration or scratching of the surface. Hard materials resist wear. Hardness is commonly measured by applying a load to a very small area of surface by means of a small piece of very hard material, such as a hard steel ball or a pointed diamond and measuring the size of the depression formed by a given load.

Softness

The reverse of hardness. Soft materials are easily shaped by cutting operations such as *turning* in a lathe, drilling etc., though other properties such as strength and ductility are also important. Many components which must resist wear must be softened to enable them to be made and must subsequently be hardened.

In some cases other properties such as thermal or electrical conductivity, magnetic properties, resistance to corrosion and appearance are important

Some commonly used materials

Materials may be grouped into a number of different types such as:

Metals, which can be further sub-divided into *ferrous* metals and *non-ferrous* metals.

Plastics, which now form a very large group having a wide range of properties.

Timber i.e. wood.

Ceramics. This term was originally applied to materials made of some kind of clay, but is now generally used for a wider range of materials.

Each type of material has special properties which make it suitable for particular uses.

2.21 Metals

Ferrous metals

These are metals in which the chief constituent is iron. Metals are seldom used in their pure state but are generally combined with other materials (usually other metals) to form *alloys* which modify the properties of the chief constituent. Possibly the most important single addition to iron is carbon which, although present only in relatively small amounts, causes important modifications in the properties of the metal.

The properties of all metals can also be influenced by *heat treatment*, which may be simply defined as controlled heating and cooling, in the solid state, with the object of obtaining particular properties in the metal.

(*i*) *Cast Iron.* As its name implies, this is iron which has been melted and poured into a suitably-shaped *mould* (usually made of sand) in which it is allowed to solidify. This is a simple and relatively inexpensive method of making complicated shapes, and one of the reasons for the use of iron in this way is the manner in which, when molten, it will flow into quite complicated moulds. Some metals cannot be satisfactorily 'cast' in this way.

Cast iron may contain up to 5% carbon, some of which is *combined* with the iron and the remainder is present in the 'free' state in the form of tiny flakes of graphite. Besides carbon there may also be small amounts of other materials, some of which are impurities which it is unduly costly to eliminate entirely while others may have been deliberately added to improve the properties of the iron. The properties may also be modified, to some extent, by suitable heat treatment.

The tensile strength is fairly low and the shear strength is about the same, but the compressive strength is about five times the tensile strength.

(*ii*) *Steel.* Steel is fundamentally an alloy of iron and carbon in which the carbon content is usually less than about 2% and is all *combined* with the iron. Steels are commonly divided into groups, according to the carbon content, approximately as follows:

Low carbon steels (mild steels), containing from	0·05–0·25% carbon
Medium carbon steels containing from	0·25–0·55% carbon
High carbon steels containing from	0·55–0·9% carbon

(it will be found that different authorities give slightly different limits for these groupings.)

Low carbon steels are the most ductile and are used for making pressed steel components. They have relatively low tensile strength (about 300–450 MN/m^2) and cannot be hardened to any useful extent by heat treatment, although there is a process, called *case hardening*, by which the carbon content of the outer skin can be increased, enabling a hard surface to be produced, the *core* remaining soft and ductile.

Medium carbon steels have a greater tensile strength (about 450–800 MN/m^2) which can be improved by suitable heat treatment. They are rather less ductile but can be hardened by heating followed by rapid cooling (quenching), though hardening reduces the ductility.

High carbon steels give the greatest hardness, though the strength may not be much greater than that of medium carbon steels. Both strength and hardness depend upon suitable heat treatment, the more severe quenching causing improvements in these directions but reducing the ductility to the point of brittleness. This brittleness can be reduced to some extent by a process called *tempering*, which follows the quenching process and consists of re-heating to a relatively low temperature and again quenching.

(*iii*) *Alloy steels.* Steels in which the distinctive properties are due mainly or solely to the carbon content and suitable heat treatment are called *carbon steels.* The addition of other alloying elements produces important modifications to some properties and steels of this kind are called *alloy steels.* I give below only a brief indication of some of the chief alloying elements.

Nickel. The addition of nickel increases the strength and toughness and amounts up to about 3·5% are often added. Larger quantities affect both the thermal expansion and the resistance to corrosion.

Chromium increases the strength and hardness. It is often used in combination with nickel to give great resistance to corrosion, one extensively used type containing about 8% nickel and 18% chromium: steels of this kind are known as *stainless steels.*

Tungsten and *Manganese* are other alloying agents, tungsten increasing the hardness and, used in conjunction with nickel and chromium and silicon form the so-called *heat resisting steels.*

Non-ferrous metals

Below are given brief details of some of the commonly used metals of this type.

(*i*) *Aluminium.* For motor vehicle work aluminium has two attractive properties. It is light, having a density less than half that of iron or steel: and it is a much better conductor of heat. In its pure state it is soft and ductile, is resistant to atmospheric corrosion, but has a low tensile strength (about 80 MN/m^2). It is seldom used in the pure state, but is usually alloyed with other metals such as copper, silicon and magnesium to produce metals with much better strength and hardness, though its resistance to corrosion is impaired.

(*ii*) *Copper* is very ductile, malleable and soft, but easily *work hardens* (i.e. becomes hard when bent or stretched). Its outstanding property is perhaps its high electrical conductivity which has led to its almost universal use for electric wiring. It was commonly used for fuel and other pipes but its relatively high cost has led to its replacement for this purpose by steel and plastics. It is resistant to corrosion and easily soldered.

(*iii*) *Brass* is an alloy of copper and zinc. It is harder than copper and is commonly used for small fittings.

(*iv*) *Bronze* is an alloy of copper and tin and has found extensive use as a bearing metal.

(*v*) *Zinc* is a soft metal having a low melting point which is easily cast. It is used for fittings such as door handles, and also for carburetter and fuel pump bodies.

2.22 Non-metallic materials

Plastics

Materials of this type become plastic above a certain temperature (which may be as low as about 100°C in some cases) and in this state they can be squeezed in dies or moulds and made into any desired shape which they retain on cooling. There are two main classes.

(*i*) *Thermosetting.* These undergo a chemical change during the moulding process and will not again become plastic on re-heating. They can, therefore, be used in conditions where they may be subjected to relatively high temperatures. They are commonly used for covers in electrical equipment (distributor caps) and for decorative covers and mouldings. They are good electrical insulators.

(*ii*) *Thermoplastic.* These can be softened by heat repeatedly so that they cannot be used at temperatures much above the boiling point of water, and some become plastic at even lower temperatures.

Many of them are transparent (celluloid, cellulose, acetate, perspex) and most can be coloured as desired by the addition of suitable pigments. Brief details of some of the commoner types are given below.

Polythene is tough, slightly flexible has a strength of about 15 MN/m^2, but becomes plastic at temperatures not much above 100°C.

Polyvinylchloride (*PVC*). This is flexible and rubber-like and is now almost universally used for insulating electrical cables. It can be coloured as desired and is completely resistant to oil and water. It is also available in sheet form and is used for upholstery.

Polytetrafluorethylene (*PTFE*) is rather similar to PVC. Its chief characteristic is that when used on itself or on most metals it has very low friction.

Nylon can be made into fibres and woven into a fabric, but can also be moulded. It is tough, has a tensile strength of about 60 MN/m^2 and low friction against metals. It is used for certain bearings and for small gear wheels (e.g. speedometer drive gears).

Timber

Wood was at one time commonly used as the framework of motor car bodies (it has even been successfully used for vehicle chassis frames) but it is now almost completely eliminated from the motor car. Almost its only use in the modern car is for decoration – e.g. instrument panels, window frames, and even here it is only found on a few makes of car.

It is still commonly used in certain types of commercial vehicle bodies, though even here it is gradually being replaced by metals.

Generally the *hard woods*, produced by slow-growing trees such as oak, ash, beech and teak are used. The chief advantages of wood are its lightness compared with metals and the ease with which it can be cut and shaped. It is, therefore, very suitable for components which are made in very small numbers where expensive machines for manufacturing would not be justified.

Hardwoods have a tensile strength of about 40 MN/m^2 but are liable to rot if not adequately protected from wet.

Ceramics

Almost the only use for materials of this kind on motor vehicles is the insulator of sparking plugs. These were at one time commonly made of a very fine natural clay called *porcelain*, but are now often made of aluminium oxide (corundum).

2.23 Surface protection

Many metals, particularly steel, suffer corrosion when exposed to the atmosphere. The corrosion usually takes the form of oxidation of the exposed surface: in the case of steel the iron oxide formed in this way is called *rust*. Corrosion of most metals is accelerated by exposure to dampness which, in the case of a motor vehicle is quite unavoidable.

Salt used by local authorities to melt ice and snow on the roads can be very corrosive to a vehicle. Water and salt together form a electrolyte, so if this solution connects two dissimilar metals then an electric battery is formed. An electrical current flowing in this way causes the metal to corrode rapidly.

Most materials are given protection from corrosion by coating their surfaces with a layer of some suitable material such as paint, or by an electro-plated coating of some corrosion-resistant metal such as chromium.

Many paints can be softened and ruined by substances such as brake fluid, acid or other general liquids used in a garage.

The frame and body structure of most cars is given a preliminary treatment such as *phosphating*. This consists of immersing the whole structure in a tank containing a liquid incorporating phosphoric acid, which converts the surface of the steel into iron phosphate. Not only does this, in itself, give some protection from corrosion but it also provides a surface to which paint will adhere strongly.

The underparts of a vehicle, which are liable to be abraded by grit and stones thrown up by the wheels, are usually given protection by a treatment known as *undersealing*. This takes the form of a sprayed-on coating of a rubber based 'paint'.

2.24 Metal joining —soft soldering

Solder is an alloy consisting mainly of tin and lead which has a melting point about 200°C. The low melting point of solder makes it most suitable for joining materials such as steel, copper and brass in situations which are not subject to excessive heat or vibration and where great strength is not required.

Fig. 2.24.1 shows the main equipment required to join two pieces of thin steel sheet. The essential items are:

Soldering iron – the means for preheating the 'work' and melting the solder.

Flux – a substance to keep the surfaces chemically clean after dirt, grease, paint and oxide has been removed by mechanical methods, i.e. file, scraper, emery cloth, etc.

Solder – the lead-tin alloy which forms the metallic bond.

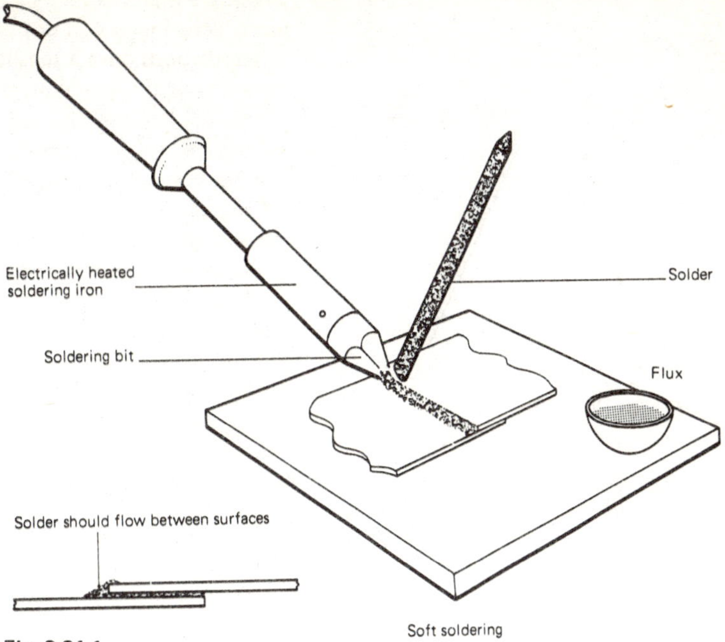

Fig. 2.24.1

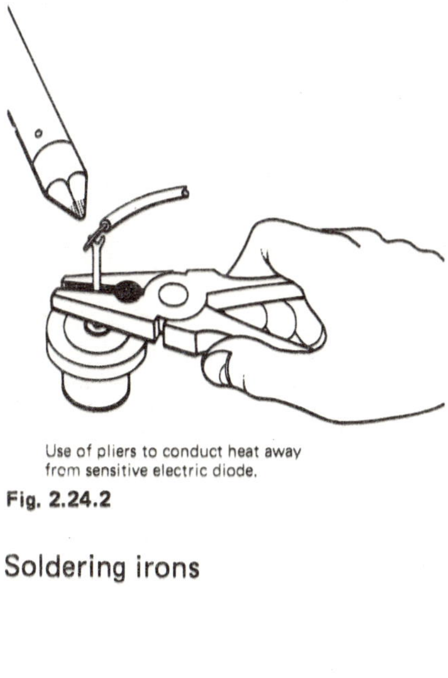

Use of pliers to conduct heat away from sensitive electric diode.

Fig. 2.24.2

Large heat loss by conduction must be avoided if the metal in the region of the join is to reach the temperature at which the solder melts. For the job shown in Fig. 2.24.1 asbestos or wood is suitable to use as a heat insulator.

When the item to be soldered forms part of either a large area or large mass, preheating by a flame is often necessary to compensate for conduction within the material. On the other hand electrical devices such as diodes are damaged by heat, so the method used in these special cases is to conduct the heat away before it reaches the sensitive regions by using pliers as shown in Fig. 2.24.2.

Referring back to the enlarged view of the joint shows that the solder has crept between the surfaces of the parent metal. This is called *capillary flow* and is necessary where strength is required. An overlap or seam joint as shown in Fig. 2.24.1 will be weak if:

(a) edges only are soldered

(b) thickness of solder between the metal sheets is not kept to a minimum.

Soldering irons

The main part of a soldering iron is a mass of copper called a 'bit', which is generally heated by an electric element. Copper is a material capable of 'storing' a large quantity of heat for its size and also it is able to conduct heat easily – these two properties make it a suitable material for a soldering bit.

Flux

Many materials absorb oxygen from the atmosphere and this process is increased when the material is heated. The oxide 'skin' acts as a barrier between the solder and the metal, thereby preventing a metal-to-metal join. To prevent this happening a substance called *flux* is applied to the metal immediately after the surfaces have been cleaned with emery or similar means.

On heating the metal, the flux coating prevents the air from contacting the metal and so keeps the surface 'chemically clean'. When the solder melts it flows under the layer of flux and on to the 'oxide-free' surface.

The tip of the soldering iron bit must also be free from oxide if the heat and solder is to flow freely from the iron. A practice called 'tinning' is performed on the iron before commencing work. The tip is cleaned with an old file, coated with flux and then heated. As the temperature of the bit increases, solder is applied to the iron until the tip is completely covered.

To perform its task the flux must:

(a) withstand the heat – it must not fully vaporise below the soldering temperature.
(b) be suitable for the metal being soldered.
(c) be suitable for the job being soldered – some fluxes are corrosive so they are unsuitable for certain jobs, e.g. electrical work requires a resin flux.

The most common fluxes which are suitable for general work are:

(a) *Zinc chloride (killed spirits)* – made by dissolving zinc in hydrochloric acid (spirits of salts).
POISONOUS to human beings. Corrosive – job must be washed in water after soldering.
(b) *Resin* – supplied as paste or in the core of the solder stock. Particularly suitable for electrical work.
(c) *Proprietary brands* – easily obtained since they are ready for use.

Other fluxes are:

Phosphoric acid	– for stainless steel
Sal-ammoniac	– for copper
Hydrochloric acid	– for zinc
Tallow	– for lead

Solders

Soft solder is an alloy of tin and lead with the addition of small amounts of other metals such as antimony and copper. Two examples are given in the Table and for reasons of simplicity only tin and lead quantities are shown

General name	*Tin %*	*Lead %*	*Melting range* °C	*Uses*
Tinmans	50	50	185 – 215	General work
Plumbers	30	70	185 – 250	Solder for filling bodies.

The melting range given represents the temperature range over which the solder is pasty; it is completely solid at the lower temperature and completely liquid at the higher temperature.

The rules to follow when soft soldering are:

(a) Thoroughly clean the surfaces to be joined.
(b) Use the correct flux.
(c) Apply adequate heat.
(d) Use the correct solder.
(e) Clean the surfaces to remove flux.

Exercises 2.25

1. After being deformed by an applied load a material returns to its original size and shape. Materials which behave in this manner are called

(a) elastic | (c) plastic
(b) ductile | (d) malleable

2. A material which can deform a considerable amount and then retain this deformation when the load is removed is called

(a) elastic | (c) brittle
(b) ductile | (d) tough

3. A material which can change its shape a large amount without breakage when hammered or rolled is called

(a) elastic | (c) malleable
(b) tough | (d) brittle

4. A material which breaks with little or no deformation is called

(a) ductile | (c) elastic
(b) tough | (d) brittle

5. A material which resists penetration or scratching of its surface is called

(a) ductile | (c) malleable
(b) hard | (d) brittle

6. A ferrous metal is one which contains

(a) iron | (c) steel
(b) carbon | (d) copper

7. When two or more metals are combined, the metal formed is called

(a) non-ferrous | (c) brass
(b) an alloy | (d) bronze

8. Which one of the following metals applies to the statement: 'the metal is brittle and is often used for exhaust manifolds'?

(a) Low carbon steel | (c) High carbon steel
(b) Medium carbon steel | (d) Cast iron

9. Which one of the following has the highest carbon content?

(a) Alloy steel | (c) High carbon steel
(b) Cast iron | (d) Mild steel

10. Which one of the following has tiny flakes of graphite present in the composition of the material?

(a) Alloy steel | (c) Low carbon steel
(b) Cast iron | (d) Mild steel

11. What type of heat treatment is carried out on low carbon steel to alter its properties?

(a) Case hardening | (c) Quenching
(b) Hardening and tempering | (d) Rapid cooling

12. Which one of the following materials is most suitable for use as a general purpose nut and bolt?

(a) Low carbon steel | (c) Cast iron
(b) High carbon steel | (d) Iron

13. Which one of the following has the lowest density?
(a) Aluminium
(b) Brass
(c) Copper
(d) Steel

14. Which one of the following has the highest thermal conductivity and highest thermal expansion?
(a) Aluminium
(b) Brass
(c) Cast iron
(d) Steel

15. Which one of the following metals applies to the statement: 'the metal is ductile, work hardens easily and is normally used for electric wiring?
(a) Brass
(b) Copper
(c) Steel
(d) Zinc

16. Which one of the following is a most suitable material for use as a bearing metal.
(a) Brass
(b) Bronze
(c) Cast iron
(d) Zinc

17. A plastics material which can be softened by heat is called
(a) thermosetting
(b) thermoplastic
(c) thermosoft
(d) thermohard

18. The insulator of a sparking plug is often made of material called
(a) PVC
(b) PTFE
(c) nylon
(d) ceramic

19. A steel has a surface which consists of iron oxide. This means that the surface has
(a) been painted with an iron oxide paint
(b) been electro-plated
(c) been phosphated
(d) rusted

20. Solder is an alloy consisting mainly of
(a) lead and tin
(b) zinc and tin
(c) lead and zinc
(d) tin and copper

21. The melting point of solder is about
(a) 100°C (b) 200°C (c) 300°C (d) 400°C

22. Before starting a soldering job the tip of the soldering bit should be cleaned, fluxed and coated with solder. This operation is called
(a) tinning
(b) tipping
(c) coating
(d) de-oxidizing

23. During soldering work the formation of an oxide film on the surface of the metal is resisted by
(a) using adequate heat
(b) occasionally scraping the surface
(c) the action of a suitable flux
(d) the metal in the solder

24. Which one of the following is a suitable flux when soldering electrical cables?
(a) Resin
(b) Zinc chloride
(c) Killed spirits
(d) Phosphoric acid

25. The composition of a typical 'tinmans' solder is about

(a) 20% tin, 80% lead
(b) 30% tin, 70% lead
(c) 50% tin, 50% lead
(d) 70% tin, 30% lead

2 SCIENCE (d) Electricity

2.30 Basic principles

Today, one grows up accepting electricity as an everyday thing – in fact the home would look very primitive if all the electrical appliances in it were removed. Electricity was first used domestically for lighting, but nowadays its use has spread to include many other applications. Each year sees the introduction of new items, which must be connected to the 'electric supply', or to use the more common term, 'mains', in order to make the item function.

In many ways the motor vehicle has a similar history to the home, as regards the use of electricity. Early motor cars only used it for producing a spark at the plug, but think for a moment about the modern motor vehicle. The number of electrical items it contains make a formidable list, and every year sees the addition of more units. Many electrical devices operate mechanical components, and in order to diagnose and rectify faults, it is essential that the mechanic should understand the basic behaviour of electricity. It is expected that you already know some of the simple rules, but let us 'spring-clean' our knowledge by covering the various aspects step by step.

Many mechanics tend to regard electricity as a mystery. One of the reasons for this attitude is that electrical current cannot be seen and only its effects are normally apparent. In order to avoid feeling this way it is recommended that you conduct as many experiments as time allows.

The electrical circuit

Let us consider wiring-up a lamp to a battery. Examination of the lamp shows that it consists of a short-length of fine wire called a filament: each end of this is connected to a soldered contact formed in the brass cap. The holder for this lamp enables a wire to connect with each end of the filament. When the ends of the two wires from the lamp are connected to the terminals of a battery, the lamp lights. This means that you have formed an *electrical circuit* – a continuous electrical path. To trace this path start at any point in the circuit and follow the path around the pattern of wire until you return to the point of origin.

You will discover that when a wire is disconnected in the circuit you have formed the light is extinguished. Fitting a switch has the same effect. It breaks or interrupts the circuit when the light is not required. If you remove the cover of a simple switch you will see how it operates mechanically.

Circuits can be illustrated in various ways; Fig. 2.30.1a shows a pictorial layout. To save time and space it is usual to represent the various electrical items by symbols. Some of the common items are shown below.

Fig. 2.30.1b shows, in theoretical form, a similar circuit to the one shown in Fig. 2.30.1a. The position of the switch indicates that the circuit is broken or *open*, so no electrical current can flow.

A flow of electricity cannot take place unless the circuit is closed or complete.

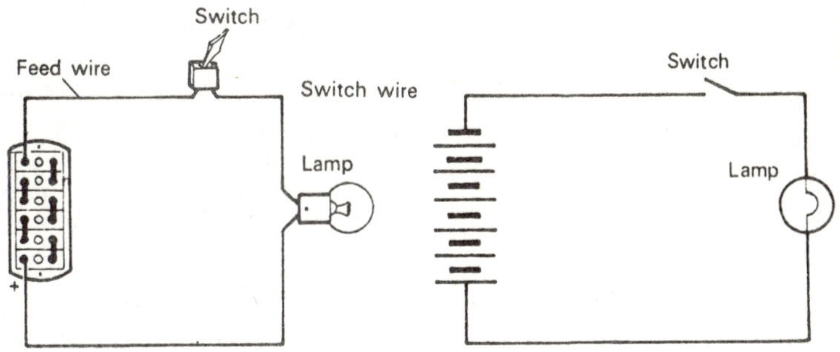

(a) (b)

Flow of electricity

Matter is made up from tiny particles called molecules and these, in turn, are built up from even smaller particles called atoms. This sub-division does not end here, because each atom consists of a large number of extremely minute particles called *electrons*, which are orbiting, at high speed, around a fixed central particle called the nucleus.

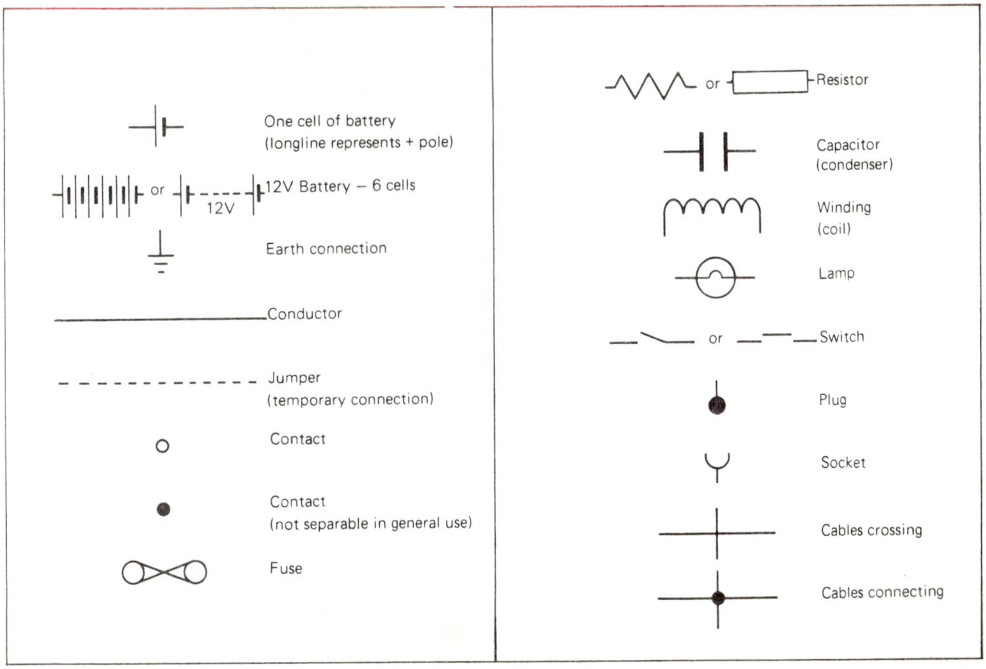

If a battery is connected to the ends of a piece of copper wire, electrons from one atom are passed to the next atom, which in turn passes on some of its electrons. From this it is seen that the closing of the circuit allows the electrons to flow around the whole circuit, including within the battery.

In some ways electricity can be likened to water – electron flow in a wire is similar to water flow in a pipe. To produce the movement of the water, some form of pump or 'forcing' device is required. This gives the necessary pressure to urge the water through the system. A battery in an electrical circuit ~~does~~ a similar job – the electrical 'pressure' between the two poles of the battery urges the electrons around the circuit.

The higher the 'pressure' of the battery, the greater will be the flow of electrons. The flow of electrons is called an *electric current* and the direction of the current in a circuit is governed by the 'polarity' of the supply. This means that when the battery connections are reversed, the direction of flow of the current is also reversed.

Each battery terminal (or pole) is given a name, positive or negative (represented by the symbols + and –). These names were selected many years ago, since it was assumed that the flow of the current, in the circuit external to the battery, took place in the direction of positive to negative. More recent discoveries show that movement of electrons is actually from negative to positive.

Conductors and insulators

It has already been shown that when a battery is connected to the ends of a piece of copper wire an electric current flows. This would not be the case if the copper wire were replaced by a length of rubber. A material which freely allows the passage of an electric current is called a *conductor*, whereas a material which resists the flow of current would be called a bad conductor or *insulator*. Copper is a good conductor, whereas rubber is a good insulator.

There is no definite dividing line between a conductor and an insulator, since most materials will act as an electrical conductor if sufficient electrical pressure is applied. In other words, a perfect insulator does not exist, and neither does a perfect conductor. The electrical conductivity of various materials is shown in the following table.

Material		*Current flow for a given electrical pressure*
Conductors	Silver	1 036 000
	Copper	1 000 000
	Aluminium	637 000
	Nickel	254 000
	Iron	175 000
	Lead	82 000
	Carbon	815
Insulators	Glass	0·000 000 000 2
	Paper	0·000 000 000 002
	Porcelain	0·000 000 000 000 002
	Ebonite	0·000 000 000 000 000 2
	Mica	0·000 000 000 000 000 02

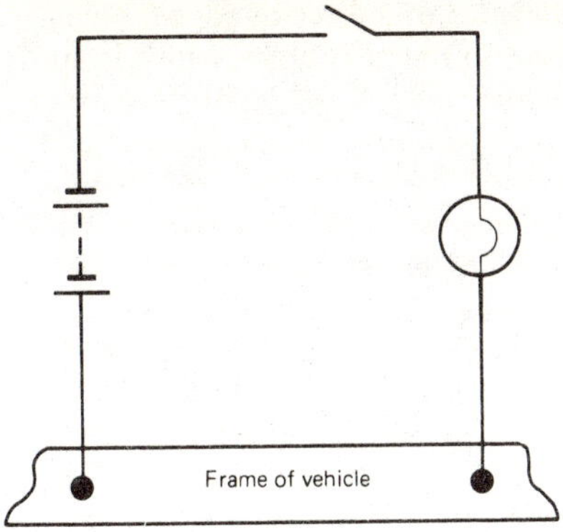

(a)

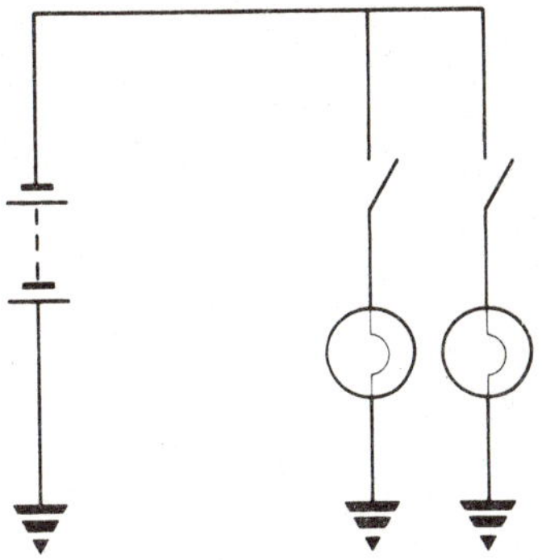

(b)

Fig. 2.30.2

The value of current flow are approximate and are only intended as a guide.

Most metals are conductors of electricity, but few can compare with copper. When this feature is linked with the fact that copper is very easily drawn into wire, it will be appreciated why the material is extensively used for electric cables on motor vehicles.

Insulated and earth return systems

The electrical circuits shown so far in this chapter have used two cables to connect the lamp to the battery. This arrangement is known as an *insulated-return* (i.r.) system, and one lead would be called the *feed* cable and the other the *return*. Since a motor vehicle incorporates a number of separate circuits, the i.r. arrangement would give a bulky, costly assembly.

On the majority of vehicles, therefore, the frame of the vehicle is used in place of one of the leads. This layout is called an *earth-return* (e.r.) system. In Fig. 2.30.2a the steel frame of the vehicle is called *earth* and in this case it would be called *positive earth*, since the positive battery terminal is connected to the frame. The system would still function if the battery terminals were reversed to give *negative earth*. In this arrangement the flow of current would be in the opposite direction, but the system is still called earth return.

In the past thirty years most British manufacturers have favoured the positive earth system, but recently many have changed to negative earth to conform with the practice in other countries.

Diagrams can be simplified by using the symbols shown in Fig. 2.30.2b. The diagram shows two lamps controlled by separate switches. Electric current flow can be controlled by the switches to produce a flow through one lamp or two.

The earth return system demands extra attention to insulation. If, for example, the cable linking the battery with the switch had a defective insulation covering which allowed the bare conductor to tough the metal frame, then an excessively high current would flow through this *short circuit*. The heating of the cable due to the high current could easily cause a fire. Since many sharp edges exist on the frame and vibration is always present when the vehicle is moving, the need for care in positioning and securing the cables is obvious.

On special vehicles such as fuel tankers, extra precautions are necessary. This type of vehicle generally uses the insulated return system for safety reasons.

Series and parallel circuits

Let us consider the wiring-up of a circuit comprising two lamps, a switch and a battery. Two basic arrangements could be used and these are known as *series* and *parallel*.

Series circuits

Fig. 2.30.3a shows the lamps connected in series, as can be seen from the fact that the lamps are connected end to end. This layout

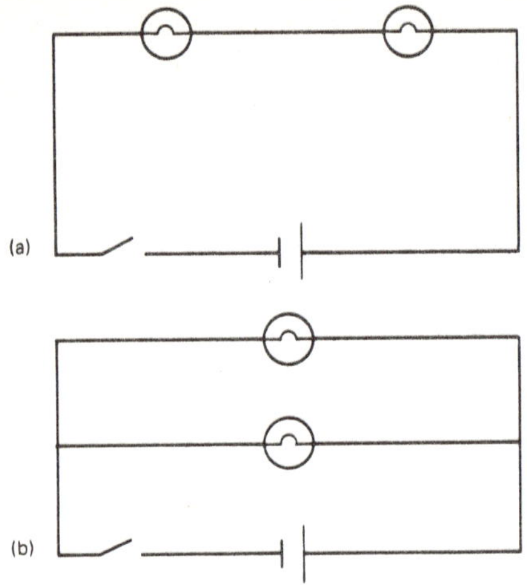

Fig. 2.30.3

means that any one electron must pass through one lamp before it reaches the other lamp. Breaking a lamp filament would interrupt the flow of current and would give a similar effect to opening the switch.

Whenever the current is passed through a material which resists its passage, the electrical pressure or potential is reduced. This point can be shown clearly by means of an actual series circuit. A circuit having one lamp only may give a bright light, but when a similar lamp is connected in series with the first, it is found that the lower electrical pressure acting on each lamp causes them both to emit only a dull light. As more lamps are inserted in the circuit, the intensity of light gradually falls until a point is reached when the lamp filament does not even glow red.

Parallel circuits

Connecting two lamps in parallel means that the lamps are connected side-by-side as shown in Fig. 2.30.3b. It must be appreciated that the expression 'side-by-side' only applies to the circuit diagram. In practice the lamps may be far removed from each other.

Lamps in parallel with each other receive the maximum electrical pressure from the battery. This means that each gives its maximum useful illumination, irrespective of the number of lamps used, assuming the battery is capable of supplying the necessary current and no other items are in the circuit. Failure of any one lamp has no effect on the others.

Parallel circuits are extensively used on motor vehicles and Fig. 2.30.4 shows a typical lighting circuit. When the switch is moved to supply current to the side and tail lamps, it will be seen that the five lamps light to full brilliance.

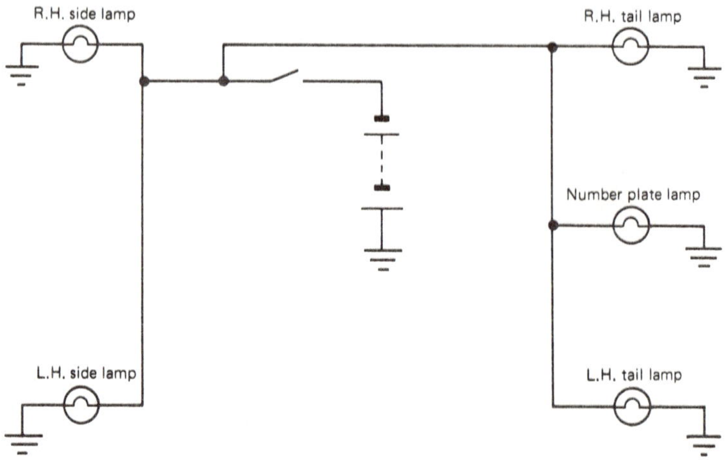

Fig. 2.30.4

2.31 Effects of an electrical current

An electric current can produce heating, magnetic and chemical effects and the motor vehicle utilises one or more of these to operate various electrical components.

Heating effect

We have already seen examples of this effect when a lamp is inserted in a circuit. During an experiment it will be seen that the wire filament of the lamp is very thin, and that when an electrical current passes, the heating effect causes the filament to become white hot. Raising the electrical pressure applied to the lamp drives a greater current through it and this would, if taken far enough, raise the temperature to a point where the filament would melt.

Magnetic effect

At some time in your life, most probably you have experimented with permanent magnets. You will remember that certain steels have the ability of retaining magnetism. The action of a magnet can be shown by placing pieces of ferrous materials (i.e. iron and steel) close to the magnet and noting the attraction between them, or by suspending a bar magnet by a length of cotton and observing the fact that it always wants to point in one direction – Magnetic North. (This is why one pole of the magnet is called North.)

You will have also seen the pattern produced after iron filings have been scattered on a sheet of paper held over a magnet. This experiment shows the presence of a magnetic field. These investigations generally lead up to the law that

Like poles repel and unlike poles attract

This law indicates that when two North poles are brought together a force is created which tends to push the magnets apart, whereas a north pole of one magnet placed near a South pole of another causes the magnets to be drawn together.

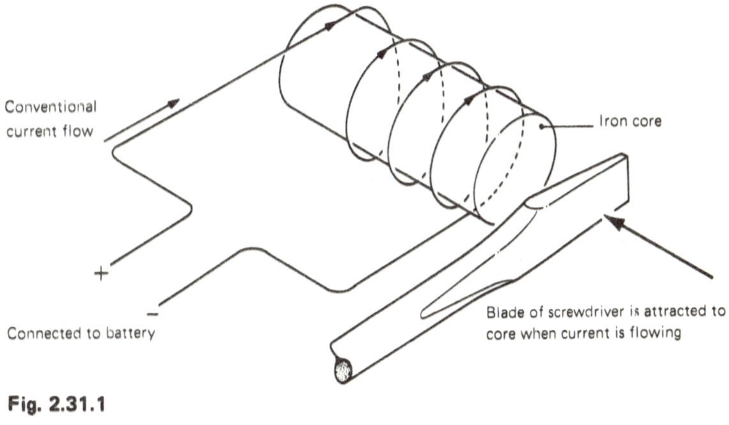

Fig. 2.31.1

A magnet can also be produced electrically. This is demonstrated by the apparatus shown in Fig. 2.31.1. A coil of wire is wound around a soft iron core and is connected to a battery. Current flow in the coil creates a magnetic field which causes the iron to become a magnet, but on disconnecting the battery the core quickly loses most of its magnetism. If a steel core were used instead of soft iron, it would retain a large amount of magnetism. Residual magnetism of this order is generally undesirable because the unit operated by this magnet should only function when the circuit is energised.

Motor vehicles utilise the magnetic effect in the following ways (the application of each is shown is brackets).

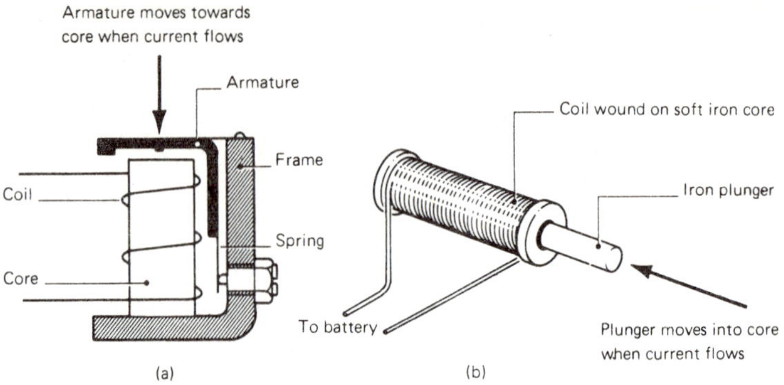

Fig. 2.31.2

Fig. 2.31.2a shows an L-shaped frame on to which is hinged an inverted L-shaped piece of soft iron called an armature. When current flows, the core becomes energised (i.e. becomes a magnet) and the armature is attracted to the core. (Cut-out, regulator, horn relays.)

The application shown in Fig. 2.31.2b uses a tubular soft iron core into which is inserted a sliding, soft iron plunger. As current flows around the coil, the core becomes a magnet and the plunger is attracted to it. This arrangement is called a *solenoid*. (Electric petrol pumps and starter switches.)

Magnetic effects also form the basis for the operation of the dynamo and starter.

Chemical effect

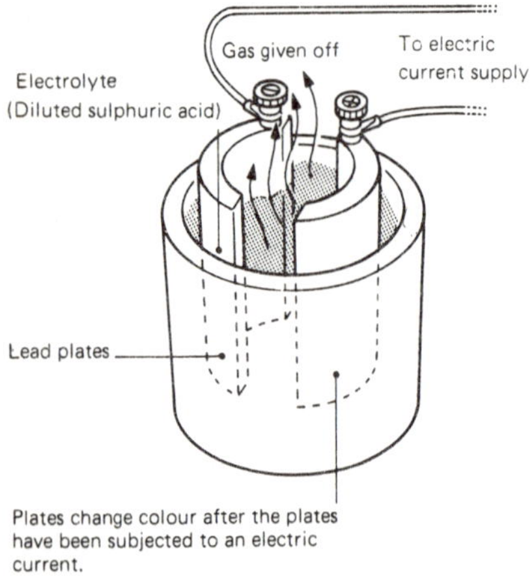

Fig. 2.31.3

This effect is demonstrated by the apparatus shown in Fig. 2.31.3. An electric current is supplied to two strips of lead immersed in a jar containing a solution of sulphuric acid (H_2SO_4) and distilled water (H_2O).

It is found that the acid solution acts as an electrical conductor and allows current to flow around the circuit. As the experiment proceeds, the lead plates change colour which indicates that a chemical change is taking place in the material of the plates. The acid solution also undergoes a change, but this cannot be seen until the current has been flowing for a considerable time. After this period it will be observed that gas gubbles are given off. If this gas is collected some of it will be found to be hydrogen, which is highly inflammable.

The chemical symbols for the acid shows that the substance consists of hydrogen, sulphur and oxygen. Passage of the electric current has caused the solution to decompose (separate). A compound which can be decomposed in this manner is called an *electrolyte*.

The chemical action just described is similar to that which takes place in a battery when it is being 'charged'.

2.32 Electrical units

Ampere (A)

The ampere, sometimes shortened to amp, is the unit used to express the rate of current flow. An *ammeter* is the instrument employed to measure current flow and this should be connected in *series* with the circuit in which the current is to be measured (Fig. 2.32.1a).

Volt (V)

The volt is the unit of electrical pressure or potential difference (p.d.). A voltmeter is the instrument used to measure the p.d. at any part of the circuit. Fig. 2.32.1b shows the method of connecting the meter when either the p.d. of the battery, or p.d. at the lamp, is required. (Sometimes the term *voltage* is used instead of p.d.) The diagram shows that the meter is connected in parallel with the part of the circuit for which the p.d. is to be measured. Mounting the meter in this manner is termed *shunting*, since the instrument's internal circuit is forming a shunt, or parallel path for the current.

A voltmeter allows only a very small current to pass through its own circuit, whereas an ammeter will pass a larger current. Connecting an ammeter in a circuit in the manner shown for a voltmeter would destroy the instrument.

Watt (W)

The watt is the unit of electric power. One unit of power is produced when a p.d. of one volt causes a current flow of one ampere. In other words,

power = p.d. x current

or watts = volts x ampere

Electrical power can be related to mechanical power.

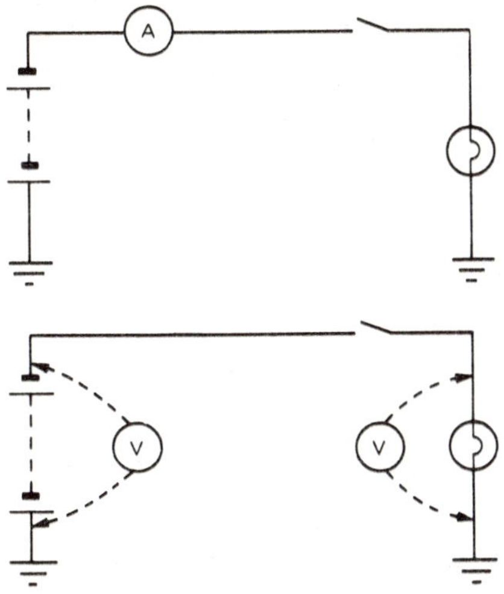

Fig. 2.32.1

Example 1

A p.d. of 12 V produces a current flow of 3 A through a certain circuit. Find the power consumed.

$$\text{Power} = \text{voltage} \times \text{current} = 12 \times 3 = 36\ \text{W}$$

(W is the abbreviation for watts.)

Example 2

A lamp is rated 12 V 36 W. Find the current which would flow through this lamp when it is connected to a 12-V supply (i.e. a supply providing a total p.d. of 12 V).

$$\text{Power} = \text{voltage} \times \text{current}$$

$$\text{Current} = \frac{\text{power}}{\text{voltage}} = \frac{36}{12} = 3\ \text{A}$$

Resistance

Opposition to the flow of electricity is termed *resistance*. This feature is the direct opposite to the term conductivity; i.e. a material which has a high electrical conductivity has a low resistance and vice versa.

The resistance of a conductor is proportional to its length and inversely proportional to its cross-sectional area. This means that if a given p.d. produces a current flow of 4 A through a cable 100 m long, then only 2 A would flow through a cable of similar diameter and 200 m long. By doubling the cross-sectional area of this longer

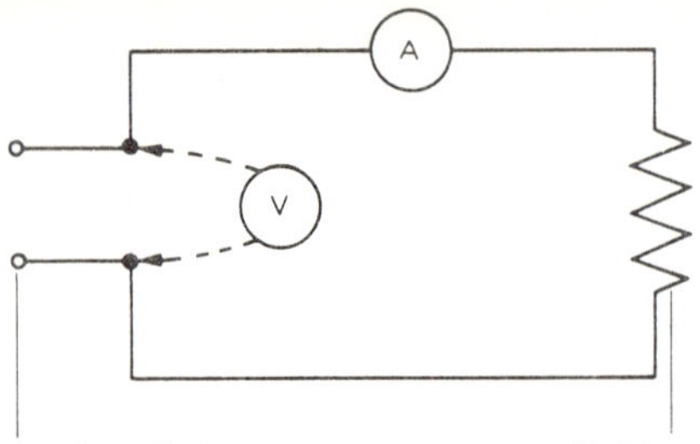

Fig. 2.32.2

cable, the resistance would be halved and the current flow would be restored to 4 A.

Many circuits incorporate a special conductor to restrict the current flow to a given value. When the conductor is selected mainly for its resistance, it is called a *resistor*.

Fig. 2.32.2 shows the layout of an experiment whereby a resistor is connected to a supply voltage, which is capable of being varied. The results are as follows:

Supply voltage	*Current (A)*	*Voltage/current*
2	1	2
4	2	2
6	3	2
8	4	2
10	5	2

It will be noted that the ratio of voltage applied to current flow produces a constant for this circuit, and in this case gives a value of 2 units. Applying this method to a number of different circuits gives similar results, so it may be concluded that the ratio $\frac{\text{voltage}}{\text{current}}$ indicates the resistance of the circuit.

The unit of resistance is called the *ohm*, and the relationship between the electrical units may be stated thus:

A p.d. of 1 V, produces a current flow of 1 A, through a circuit of resistance 1 ohm.

Applying this standard to the experiment shown in Fig. 2.32.2, indicates that the resistance of the circuit is 2 ohms.

The resistance of any circuit can be found by either using the expression

$$\text{Resistance} = \frac{\text{voltage}}{\text{current}}$$

or by using an instrument called an *ohmmeter*.

Resistance of a material is affected by temperature. In most metals metals a rise in temperature causes the resistance to increase, but the change is not very great unless the temperature variation is large.

Summary of electrical units

Term	*Symbol*	*Definition*
ampere	A	The unit of current flow
volt	V	The unit of potential difference or electrical pressure
watt	W	The unit of power (watt = volt x ampere)
ohm	Ω	The unit of resistance

Metric units are often used as a prefix to the above terms:

kilo meaning 1000

milli meaning $\frac{1}{1000}$ OR 0·001

micro meaning $\frac{1}{1\,000\,000}$ OR 0·000 001

therefore

$$3\ \text{kW} = 3\ \text{kilowatt} = 3000\ \text{W}$$
$$9\ \text{kW} = 9\ \text{kilovolt} = 9000\ \text{V}$$
$$4\ \text{mV} = 4\ \text{millivolt} = \frac{4}{1000}\ \text{V}\quad \text{OR}\quad 0{\cdot}04\ \text{V}$$
$$7\ \mu\text{V} = 7\ \text{microvolt} = \frac{7}{1\,000\,000}\ \text{V}\quad \text{OR}\quad 0{\cdot}000\,007\ \text{V}$$

2.33 Relation between heating effect and current

It has already been stated that the generation of heat is one effect of an electric current. When a current passes through a conductor, heat is generated.

The number of heat units generated in a conductor depends on:

1. the resistance of the conductor,
2. the current flowing,
3. the time during which the current flows.

If the conductor is of suitable size, this heat will pass to the air and the temperature of the conductor will not rise appreciably. If, however, the current is increased for example, then the conductor may overheat and melt. Even if the conductor did not melt, the heat would be sufficient to destroy any insulating covering and perhaps cause a fire.

This fact can be usefully employed to protect a circuit and provides the principle of a *fuse*. Generally made of a lead/tin alloy, the fuse consists of a thin piece of wire with a small surface area to dissipate the heat. When the current exceeds a given value, therefore, the fuse will melt and the circuit will be broken.

Exercises 2.34

1. The symbol for a battery cell is 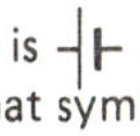 . What polarity is represented by the longer line and what symbol is used to show this?

(a) Positive which is shown as +
(b) Positive which is shown as –
(c) Negative which is shown as +
(d) Negative which is shown as –

2. The symbol  represents a:

(a) coil (c) capacitor
(b) lamp (d) switch

3. The symbol  represents a:

(a) coil (c) capacitor
(b) lamp (d) switch

4. The symbol  represents a:

(a) coil (c) capacitor
(b) lamp (d) switch

5. The symbol  represents:
(a) a resistor (c) a coil
(b) an earth connection (d) a switch

6. The symbol  represents:
(a) a resistor (c) a plug
(b) an earth connection (d) a cell

7. A material which freely allows the passage of an electric current is called:
(a) an insulator (c) a capacitor
(b) an isolator (d) a conductor

8. A material which resists the flow of an electric current is called:
(a) an insulator (c) a capacitor
(b) an isolator (d) a conductor

9. Which one of the following materials is the best electrical insulator?
(a) Copper (c) Carbon
(b) Lead (d) Porcelain

10. Which one of the following materials is the best electrical conductor?
(a) Copper (c) Carbon
(b) Lead (d) Porcelain

11. As applied to electrical circuits the abbreviation 'e.r.' is commonly used to represent:
(a) electrical resistance (c) earth return
(b) extra reach (d) each resistor

12. An 'open' circuit means that:
(a) one part of the circuit is earthed
(b) a fuse is fitted in the circuit
(c) the circuit is incomplete
(d) part of the circuit is exposed

13. The frame of a vehicle cuts through the covering of a cable and causes a large current to flow to 'earth'. This fault is normally called:
(a) a frame circuit (c) an open circuit
(b) a fused circuit (d) a short circuit

14. The purpose of a switch in a lighting system is to:
(a) open the circuit when the lights are required
(b) open the circuit when the lights are not required
(c) short the circuit when the lights are required
(d) short the circuit when the lights are not required.

15. A battery in a circuit causes tiny particles to move from one atom to the next. These particles are called:
(a) molecules (c) volts
(b) electrons (d) amps

16. The term 'negative earth' means that the negative battery terminal is
(a) not used (c) connected to the road
(b) insulated from the frame (d) connected to the frame

17. When is an insulated return system used?
(a) On the majority of popular cars
(b) On the majority of commercial vehicles
(c) On vehicles operating in wet climates
(d) On vehicles having a high fire risk.

18. Two lamps are connected to a battery in an arrangement which causes the total current supplied by the battery to pass through each lamp. This circuit is called:
(a) series
(b) parallel
(c) earth return
(d) insulated return

19. Two 12 V 6 W lamps are connected to a battery in an arrangement which causes each lamp to pass half of the total current supplied by the battery. This circuit is called:
(a) series
(b) parallel
(c) earth return
(d) insulated return

20. Two lamps are connected in series with each other. What is the effect if one lamp filament breaks? The 'good' lamp will:
(a) become brighter
(b) become dimmer
(c) operate normally
(d) not light

21. Two lamps are connected in parallel with each other. What is the effect if one lamp filament breaks? The 'good' lamp will:
(a) become brighter
(b) become dimmer
(c) operate normally
(d) not light

22. What arrangement is used to connect the various lamps in the side-and-tail lamp circuit of a motor vehicle. Lamps are connected in:
(a) series
(b) parallel
(c) line
(d) polarity

23. Heating is one of the three effects of an electric current. The other two effects are:
(a) magnetic and thermal
(b) thermal and mechanical
(c) mechanical and chemical
(d) chemical and magnetic

24. Soft iron is often used as a core for an electro-magnet because it
(a) is not affected by a magnetic field
(b) does not retain a large amount of magnetism
(c) is a non magnetic material
(d) keeps cool.

25. Which words makes the following statement correct:
'like magnetic poles . . . . , unlike poles . . . . ?
(a) attract . . . .resist
(b) resist . . . . repel
(c) attract . . . . repel
(d) repel . . . . attract

26. What is an electrolyte?
A solution which is:
(a) chemically changed by electric current
(b) used as a battery plate
(c) used as a battery separator
(d) electrically charged distilled water.

27. The unit of electrical current is the:
(a) ampere (b) ohm (c) watt (d) volt

28. The unit of electrical potential difference is the:
(a) ampere (b) ohm (c) watt (d) volt

29. The unit of electrical power is the:
(a) ampere (b) ohm (c) watt (d) volt

30. The unit of electrical resistance is the:
(a) ampere (b) ohm (c) watt (d) volt

31. Electrical 'pressure' in a circuit is measured by:
(a) an ammeter (c) a wattmeter
(b) an ohmmeter (d) a voltmeter

32. The 'charge' supplied by a generator is normally measured by:
(a) an ammeter (c) a voltmeter
(b) a hydrometer (d) a wattmeter

33. The power consumed when a p.d. of 12 V produces a current flow of 9 A is:
(a) 0·75 W (b) 1·3 W (c) 21 W (d) 108 W

34. The standard current flow through a lamp rated at 12 V 60 W is:
(a) 0·2 A (b) 5 A (c) 48 A (d) 720 A

35. An instrument shows a reading of 2 mV. The unit is a:
(a) megavolt (c) microvolt
(b) millivolt (d) metrevolt

36. A fuse is fitted in an electric circuit to:
(a) prevent a short circuit (c) protect the circuit
(b) prevent an open circuit (d) protect the lamps.

2.35 Rating of lamps

Lamps are rated by their voltage and wattage, e.g. 12 V 48 W. Power, expressed in watts, is the product of voltage and current, so it is a simple matter to determine the current which a given lamp will pass. In the case of the 12 V 48 W lamp, the current flow will be 4 A.

The following table shows a typical set of lamps for a vehicle using a 12-V system.

Head (Main & dip)	60/45 W	Ignition warning	2·2 W
Side	6 W	Main beam warning	2·2 W
Flasher	21 W	Oil warning	2·2 W
Tail and stop	6/21 W	Interior	6 W
Number plate	6 W		

Current loading of a given circuit should be known when a new cable has to be fitted to a vehicle for which the recommendations are unknown. If the length is not excessive, a current flow of $\frac{1}{2}$ A per strand of 0·25 mm cable is suggested.

Example

A new cable is to be fitted to supply two headlamps of rating 12 V 36 W. Determine the cable size required.

12 V 36 W gives a current flow of 3 A.

Since the lamps are connected in parallel, then the total current flow = 3 + 3 = 6 A.

Number of strands of 0·25 mm cable required:

$$= \frac{\text{total current flow in circuit}}{\text{current flow allowed per strand}}$$

$$= \frac{6}{\frac{1}{2}}$$

= 12 strands

This size of cable is not generally stocked, so the next size up is selected.

2.36 Circuits

Method of connecting and switching lamps

It has already been stated that to give maximum light, lamps must be connected in parallel. Fig. 2.30.4 shows a typical side and tail lamp circuit comprising five lamps, all connected in parallel and controlled by a single switch.

The lamps in the circuit have ratings of:

side lamps	12 V 6 W
rear lamps	12 V 6 W
number plate lamp	12 V 6 W

so the switch must carry a current of $2\frac{1}{2}$ A.

Headlamps are not always required when the side lamps are being used, so a separate switch, which is normally incorporated in the main lighting switch, is fitted. Fig. 2.36.1 shows the addition of a headlamp circuit to the original side and rear lamp system. This diagram also shows an alternative method of representing the circuit. Instead of showing the items in their respective positions, the components are grouped together, and this makes the diagram easier to understand. In this diagram you will see that the current 'fed' to the headlamps, first passes a *dip* switch. This switch distributes the current to either the main or dip beam. A light on the instrument panel indicates to the driver when the main beam filament is being used.

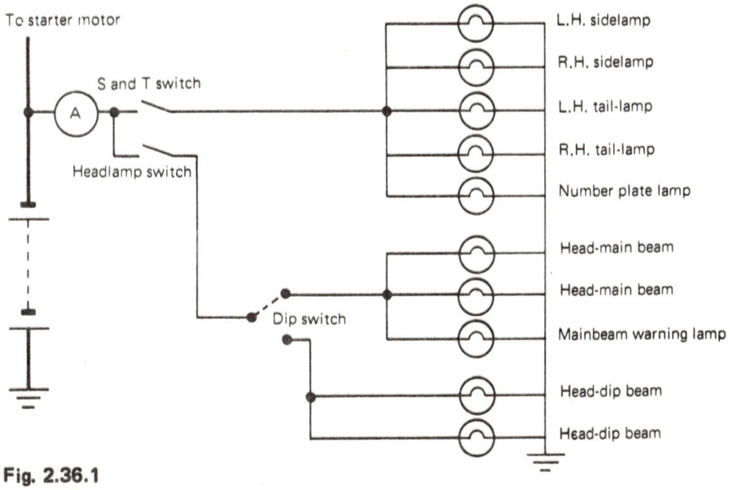

Fig. 2.36.1

Dimming arrangements

Headlamps

Headlamp circuits must include some form of anti-dazzle device. In the previous case the circuit consisted of a twin filament lamp to give dip and main beam illumination. The dip filament has a lower wattage and is so positioned in relation to the lamp reflector that the beam is deflected downwards and towards the near-side of the vehicle.

Panel lamps

The instruments are normally illuminated by two or more lamps incorporated in the fascia panel. These are generally controlled by a separate panel switch, which is 'fed' from the lamp side of the switch controlling side and rear lamps.

Sometimes this panel switch includes a dimmer control to enable the driver to vary the illumination of the panel from a dull glow to maximum brilliance. This action is provided by a variable resistor: the form shown in Fig. 2.36.2.

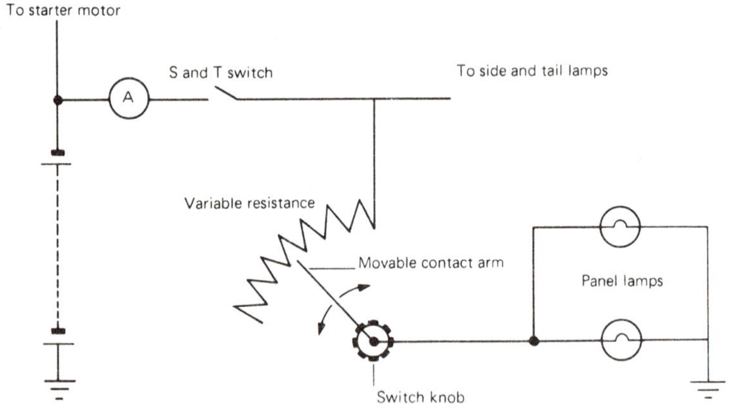

Fig. 2.36.2

A length of resistance wire (a special alloy steel wire with a much higher resistance than normal copper wire) is closely wound to form a coil. A movable contact arm, linked to the knob of the switch, wipes over the coil to vary the effective length of the resistor.

In the position shown, the arm is mid-way around the resistor, so the light will be lit accordingly. Moving the arm towards the 'off' position increases the resistance, and lowers the p.d. acting on the lamps. (This can be verified by connecting a voltmeter across the lamp.)

Coil-ignition systems

Most modern vehicles use this form of ignition and Fig. 2.36.3 shows the main components required for a 4-cylinder engine. (At this stage only a brief description is given since it will be covered in detail in Part II of the syllabus.)

The *coil* transforms the battery p.d. to the voltage required to produce a spark at the plug. This requires about 8000 V (8 kV) under normal conditions. The coil does not give a steady current at this voltage, but provides a surge only when a spark is required. This

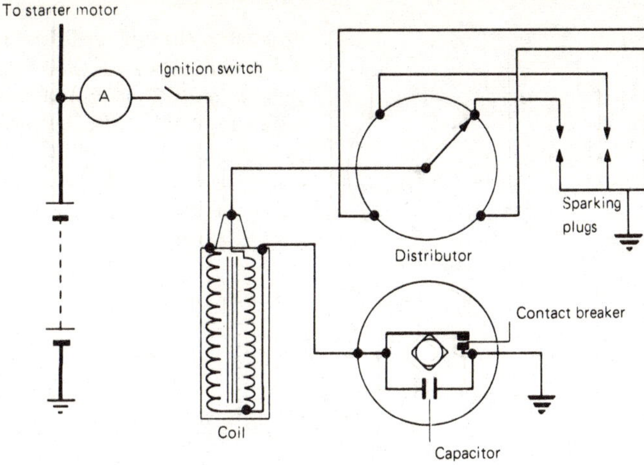

Fig. 2.36.3

action is produced when the primary circuit is momentarily broken by the *contact breaker.*

Arcing across the contacts of the breaker is prevented and more rapid interruption of the current is achieved by connecting a capacitor in parallel with the contacts.

An ignition switch controls the flow of current in the low tension or primary circuit and enables the engine to be stopped when necessary.

When the engine has more than one cylinder, the high voltage impulse must be directed by a rotary switch to the appropriate sparking plug. This switch is called a *distributor* and consists of a rotor and a series of brass segments to correspond with the number of cylinders. High-tension leads connect the distributor to the sparking plugs, and the thick insulation on these leads ensures that the h.t. charge cannot take any path to earth other than that across the spark plug electrodes.

2.37 Batteries

Electromotive force

We have seen that some device is necessary to cause a difference in potential in order for eletrical current to flow in a conductor. The chemical cell is probably the most common method which comes to mind. In this case the difference in potential between two dissimilar metal plates produces a pressure called *electromotive force* (generally shortened to e.m.f. and expressed in volts). This 'drives' the current around the circuit, and exists even when the cell is disconnected. In many ways the cell is similar to an inflated tyre – the pressure wants to escape so that it may equalise with the surrounding air, whereas the cell wants to equalise the potential in each plate.

Potential

Let us examine the meaning of the term 'potential' more closely. The word suggests 'energy stored', and in a case of the tyre quoted above, the energy is apparent. When the electrical charges contained in two dissimilar metal plates are compared, it is found that the electrical charge in one plate is greater than in the other plate. If the two plates are immersed in an electrolyte and externally connected to form a circuit, then the difference in the 'charge levels' will produce a current flow. This has the effect of lowering the potential in one plate and raising it in the other.

In an attempt to identify the electrical charges, one was termed positive and the other negative. In the early days it was assumed that the flow was in the direction of high potential (positive) to low potential (negative), but recent theories show that the electron flow is from negative to positive. However to apply the basic rules of electricity, which were introduced in the nineteenth century, it is necessary to consider that the flow takes place from *positive to negative*.

Primary cells

Chemical cells have been used for many years. In fact it was in the late eighteenth century that Volta immersed two dissimilar metals in an electrolyte and produce an electric current flow.

There are many combinations of metals and electrolytes used to produce a simple cell. One type employs a copper (+) and a zinc (–) plate immersed in a solution of sulphuric acid and water. The cell initially gives an e.m.f. of about 1·0 V, but after a short time it is seen that the zinc plate starts to dissolve and bubbles of gas form on each plate. This quickly leads to a fall-off in the output of the cell, so in this form the cell is not very efficient.

When this type of cell is exhausted – in other words, when the zinc has dissolved – it has to be replaced with new material. Since it is impossible to restore the cell by electrical means, it is called a *primary* or *irreversible cell.* (Dry type primary cells are used in many torches and portable radios.)

Secondary cells

Its limited life and the irreversibility of its action make the primary cell unsuitable for motor vehicles. Secondary type cells are used, because they allow the potential to be restored by supplying the cell with a charge current, which flows in a direction opposite to the flow given out by the cell – hence the term *reversible*.

Theoretically this type of cell should last indefinitely, but owing to mainly mechanical problems in construction, its average life seldom exceeds two or three years.

Lead-acid batteries

A battery is a number of cells connected together: 'lead-acid' refers to the materials used in the cell. This is the most common type of battery employed on motor vehciles.

Each cell consists of two sets of lead plates immersed in a solution of sulphuric acid (H_2SO_4) and distilled water: the solution is termed the *electrolyte.* A nominal p.d. of 2 V is obtained from each cell and by connecting cells in series the battery p.d. = 2 V x number of cells (Fig. 2.37.1). Mounting the cells in parallel with each other has no effect on the p.d. of the battery: it remains at 2 V irrespective of the number of cells linked together but it does alter the capacity (the period of time that the battery will supply a given current.)

Charge and discharge process

When the battery is ready for use (fully charged), the positive plate is composed of lead peroxide and the negative plate is spongy lead.

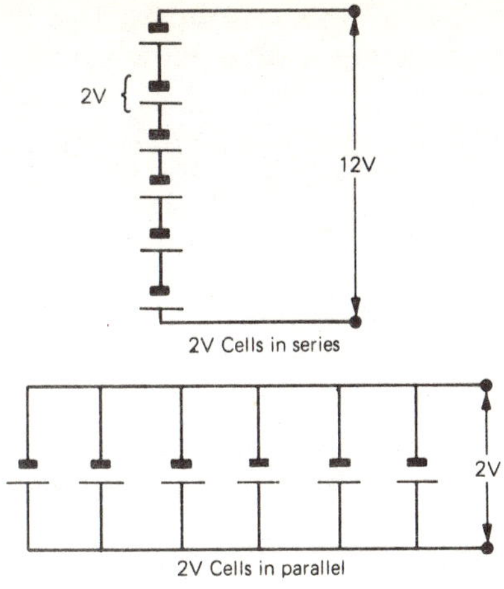

Fig. 2.37.1

Connecting the battery to an external circuit, causes a discharge current to flow. This has the effect of lowering the potential of the battery. While this flow is taking place, a chemical action called *electrolysis* is occurring within the battery. Sulphur atoms, contained in the sulphuric acid, combine with some of the lead oxide and change the plate composition to lead sulphate. As this process continues some of the acid changes to water.

Applying a charge current (a flow in the opposite direction to the discharge current) has the effect of reversing the chemical process – the plates return to lead peroxide and spongy lead and the electrolyte reverts to its original density.

The process may be shown as follows:

	Positive plate	*Electrolyte*	*Negative plate*	
↑ CHARGE	Lead peroxide	ACID + water	Spongy lead	DIS-CHARGE ↓
	Lead sulphate	acid + WATER	Lead sulphate	

Sulphation

The previous table shows that lead sulphate forms on each plate when the battery is in a discharged state. This substance resists the passage of an electric discharge current, but assuming the battery has not been discharged past the normal limit, then it is possible to 'break-down' the sulphate and restore the cell to its charged state.

To obtain the maximum life of the battery, excessive sulphation should be avoided. In addition to the cause already mentioned, sulphation also results from topping up with acid instead of distilled water.

Electrolyte composition

The relative density of the electrolyte for a new battery is generally 1·280 and this is obtained by diluting concentrated sulphuric acid with distilled water. The mixing should take place in a glass or earthenware container by *adding the acid to the water*. If this rule is broken, serious injury could occur.

The *state of charge* of the cell can be ascertained by measuring the *density* of the electrolyte. A typical table of relative density values for a lead-acid battery fitted to a motor vehicle would be:

Fully charged	1·250
Half charged	1·200
Fully discharged	1·150

Relative density (formerly called specific gravity) is:

$$\frac{\text{mass of a given volume of the substance}}{\text{mass of an equal volume of water}}$$

A relative density of 1·250 indicates that the acid electrolyte is 1·25 times as heavy as water.

Relative density values, obtained by means of a *hydrometer* (Fig. 2.37.2), show that the proportion of water to acid increases as

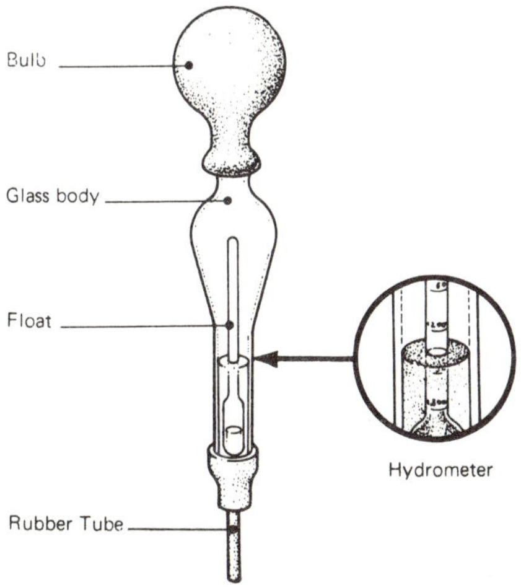

©

Fig. 2.37.2

the cell becomes discharged. As this water content increases, the temperature at which the electrolyte will freeze rises.

Battery capacity

Capacity is expressed in *ampere-hours*. This is the current (A) that a battery will supply for a given time (hours):

$$\underset{\text{(ampere-hours)}}{\text{Capacity}} = \underset{\text{(A)}}{\text{current}} \times \underset{\text{(hours)}}{\text{time}}$$

Capacity is based on either a 10- or 20-hour rate For example, a 38 ampere-hour battery (based on a 10-hour rate) indicates that it will give a current flow of 3·8 A for a period of 10 hours. Generally, the 20-hour rate gives a higher value for capacity since the task of supplying a lower discharge current is less arduous.

Cell capacity is governed by the total surface area of the plates, so a larger capacity is achieved by extending the size or increasing the number of plates.

Voltage variation

When a fully charged cell is first put on discharge, the terminal p.d. is about 2·1 V, but this quickly drops to 2·0 V, where it remains for the major portion of discharge period. Towards the end of the period, the p.d. falls more rapidly until a p.d. of 1·8 V is reached and this is regarded as the fully discharged condition. If the cell is discharged below this p.d., its life will be shortened.

Terminal p.d. during the charging process rises gradually from 2·0 V to 2·5–2·7 V when the cell is fully charged, but soon falls to about 2·1 V when the charge current is stopped. These readings represent p.d. It must be remembered that current must be flowing when the reading is taken.

The extent of the charge or discharge current considerably affects the voltage readings. This point shows up clearly when a starter motor is cranking a stiff engine. Continual operation of a starter 'drawing' a current of approx. 200 A produces a tremendous load on the battery. The voltage soon drops to the discharged value, but after a period of inactivity, the battery recovers its original voltage. A similar action occurs when a large current is supplied to the battery. After a relatively short time, the cell gives the impression of being fully charged, but if the current flow is stopped and the cell checked after a period of time, then it is found to be only partly charged.

Exercises 2.38

1. A vehicle lighting circuit consists of two 12 V 48 W headlamps and four side and tail lamps of 12 V 6 W. When all these lamps are operating with the engine stationary the vehicle ammeter should register:

(a) 2·5 A (c) 10 A
(b) 8·5 A (d) 16 A

2. Assuming the length is not excessive, the number of strands of 0·25 mm cable to carry a current of 8 A is:

(a) 2 (c) 10
(b) 8 (d) 16

3. The purpose of the capacitor in a coil-ignition system is to:
(a) break the primary winding at the appropriate time
(b) produce a 'clean' break at the contact breaker
(c) hold the charge until a spark is required
(d) prevent arcing at the rotor tip

4. High tension ignition leads are heavily insulated to:
(a) prevent the current short circuiting the plug
(b) ensure that the plug receives its high current
(c) stops water contacting the metal cable
(d) strengthens the lead against vibration

5. One essential difference between a primary and secondary cell is that a secondary cell
(a) cannot be charged
(b) is always larger
(c) contains an electrolyte
(d) is reversible

6. How many cells has a 12 V lead-acid battery?
(a) 3
(b) 6
(c) 9
(d) 12

7. What is the nominal p.d. when three 2 V lead-acid cells are connected in series?
(a) 2
(b) 3
(c) 6
(d) 12

8. Two 6 V batteries are connected positive to positive and negative to negative. The p.d. between the output terminals of this arrangement should be:
(a) 3
(b) 6
(c) 12
(d) 18

9. Examination of two lead-acid batteries shows that battery 'A' has nine plates per cell and battery 'B' seven plates per cell. This means that battery A has:
(a) a voltage of 18 V
(b) a voltage which is greater than B
(c) has a large capacity than B
(d) has a smaller capacity than B

10. Battery capacity is expressed in:
(a) ampere
(b) volts
(c) ampere-volts
(d) ampere-hours

11. The composition of the electrolyte used in a lead-acid battery is:
(a) lead peroxide and sulphur
(b) sulphur and sulphuric acid
(c) sulphuric acid and distilled water
(d) distilled water and lead peroxide

12. One plate of a lead acid battery is composed of lead-peroxide. To which plate does this apply and what is the state of charge of the battery?
(a) Positive plate and battery fully charged
(b) Positive plate and battery discharged
(c) Negative plate and battery fully charged
(d) Negative plate and battery discharged

13. What is meant by the term 'battery sulphation'?
A substance on the plates which:
(a) is applied by the manufacturer to protect the cell
(b) resists the passage of an electric current
(c) prevents the active material falling from the grid
(d) protects the cell and extends the life

14. Concentrated sulphuric acid should be diluted by:
(a) dissolving zinc chippings in the acid
(b) separating the sulphur from the acid
(c) adding water to the acid
(d) adding acid to the water

15. A lead-acid battery in a fully charged state has a relative density of about:
(a) 1·100 (c) 1·250
(b) 1·150 (d) 1·350

16. A lead-acid battery is considered 'fully discharged' when the relative density falls to:
(a) 1·100 (c) 1·250
(b) 1·150 (d) 1·350

17. The electrolyte in a battery has a relative density of 1·200. Given that 1 litre of water has a mass of 1 kg, the mass of 1 litre of battery electrolyte is:
(a) 0·08 kg (c) 1·2 kg
(b) 1 kg (d) 12 kg

18. During the major part of the discharge period, the p.d. of a lead-acid cell is:
(a) 1·8 (c) 2·2
(b) 2·0 (d) 2·4

19. When a lead-acid cell reaches its fully discharged condition the p.d. is:
(a) 1·8 (c) 2·2
(b) 2·0 (d) 2·4

20. The capacity of a battery is stated as 60 Ah on a 10 hour rate. This means that during a 10 hour period the battery will supply a current of:
(a) 6 A (c) 60 A
(b) 10 A (d) 600 A

3 DRAWING

3.1 Projection

Any manufacturer's service manual shows how widespread is the use made of drawings. Whether it indicates a form of construction or a suggested method of repair, a drawing reduces the written description and conveys ideas in a form from which the actual item can easily be recognised.

The purpose of this section of the book is to clarify the layout of an engineering drawing, and to enable a student to convey what he knows about a component to an examiner by means of drawings. Pictorial descriptions may be likened to the written word – they must be capable of being understood by some other person, so certain universally accepted rules should be obeyed. These rules are stated in British Standard 308 and they should be adopted by all engineering students.

Engineering components are three-dimensional; i.e. they possess length, breadth and depth, so the drawing should convey the detail of each side. In the majority of cases only three sides of the figure need be shown. These sides may be shown in one drawing (pictorial projection) or each side could be shown separately (orthographic projection).

The following methods of projection show the way in which a battery could be drawn.

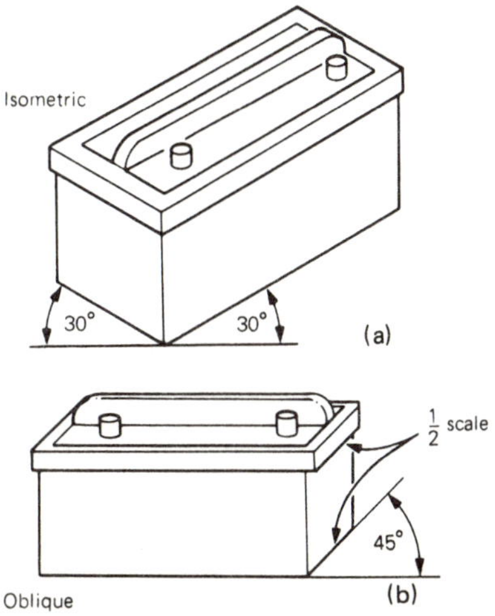

Fig. 3.1.1.

Pictorial projection

Two forms are used by engineers: (i) Isometric; (ii) Oblique.

In an *isometric* drawing the component is viewed from the corner, and lines joining the corner of the battery are inclined at 30° to the horizontal (Fig. 3.1.1a). All lines representing the length of the component are of equal length – this is the meaning of the term isometric.

In an *oblique* drawing the front face of the object is drawn and the side is projected on an axis at 45° to the horizontal. The depth of the component is drawn half scale (Fig. 3.1.1b).

Orthographic projection

This method shows a separate view of each face of the object; the number of views necessary will be governed by the shape of the object. Rectangular solids normally require three views; a front elevation, end elevation and plan. If circumstances justify it a view of the base and the other end are also shown.

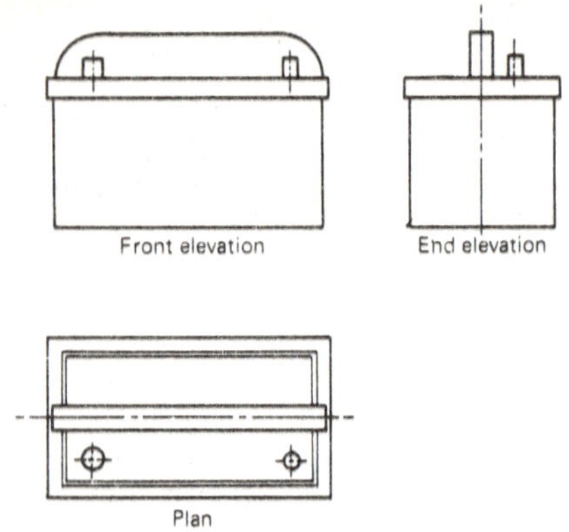

Fig. 3.1.2

The position each view occupies on the paper is controlled by the projection angle used. Fig. 3.1.2 is drawn in a form called *1st Angle*. This is the common arrangement used in Europe. The American-favoured *3rd Angle* projection system is also used in this country.

1st Angle Projection. To determine the position each elevation of the object should occupy on the paper consider the demonstration shown as Fig. 3.1.3.

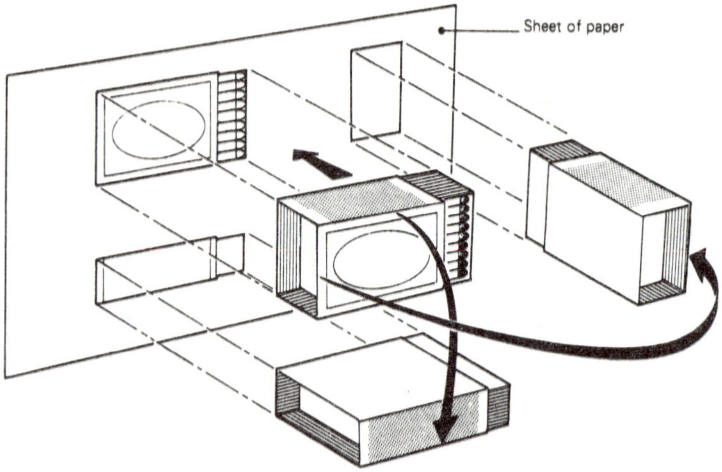

Fig. 3.1.3

Using a matchbox to represent the object, lay the box on its side and draw the face you can see (front elevation).

Place the box on this drawing and move the box to the right by rocking it through one right angle. Draw the face you can see. This drawing is positioned on the side to which the box has moved; that is, the right-hand side in this case.

Return the box to the original position and again rotate the box through one angle to move it towards the bottom of the paper. Draw the face you can see (plan).

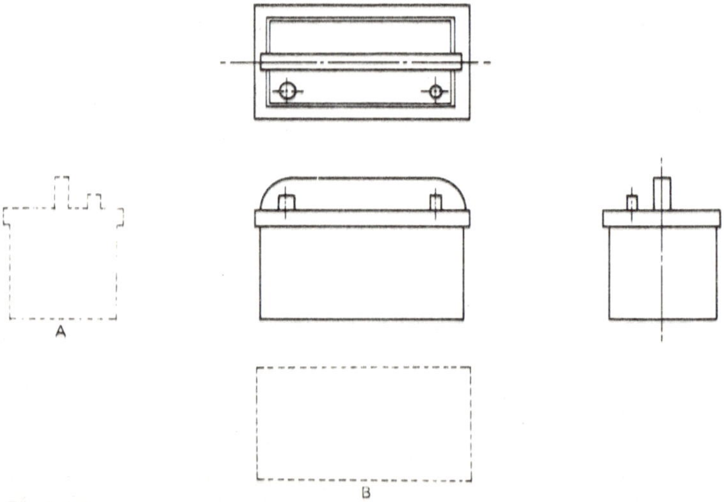

Fig. 3.1.4

To summarise:

Turn the box through *one right angle* and the position it occupies will indicate where the *1st angle* elevation should be drawn.

3rd Angle Projection. Using the same matchbox demonstration, draw the faces you can see, but rotate the box through *three right angles* to show the position for each view. In this case it will be seen that the faces occupying the plan and end elevation positions are opposite those drawn in 1st angle projection.

Fig. 3.1.4 shows a battery drawn in 3rd angle projection. Further views could be inserted in positions A and B if the three views did not enable the full details to be shown.

Types of line

B.S. recommend the use of the following types of line for general engineering drawings:

A Continuous (thick)		Visible outlines
B Continuous (thin)		Dimension lines Projection lines Hatching or sectioning
C Short dashes (thin)		Hidden details
D Long chain (thin)		Centre lines
E Long chain		Cutting or viewing planes

3.2 Sketching

Although the ability to produce an excellent sketch is a gift which is not possessed by everybody, results can be greatly improved if a few simple exercises are practised.

Having obtained a soft pencil (HB or B) and sharpened it to a conical point you should first practise drawing straight lines. The line should be produced in *one* sweep – the hand moving from left to right in the case of a right-handed person.

Once you have achieved a good straight line, try sketching simple items such as rectangular objects. At this stage attention should be given to proportion. The object should initially be sketched in faintly to obtain the correct shape and then should be lined-in to give the finished form.

Consider the production of an isometric sketch of a Vee block (Fig. 3.2.1).

Stage 1. Draw construction lines to show the corner. Each line should form an angle of 30° to the horizontal.

Stage 2. Draw the outline of the block. To ensure the lines are parallel, turn the paper so that the pencil is always moving from left to right.

Stage 3. Draw in detail of the block and complete the sketch by lining-in the required portion.

Curves and circles sometimes cause difficulty. If you cannot draw a satisfactory circle freehand, then use some type of aid. Two of the

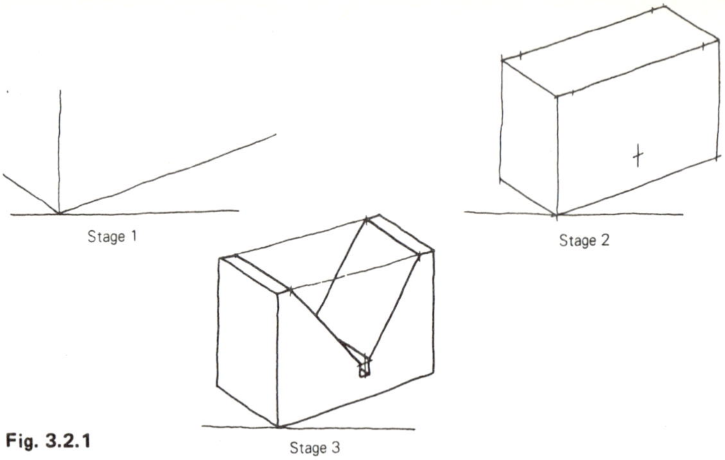

Fig. 3.2.1

many techniques of freehand drawing a circle are shown in Fig. 3.2.2.

Whenever a curve is drawn it is advisable to use the sweep of the wrist. The pencil should not be pushed away from the body.

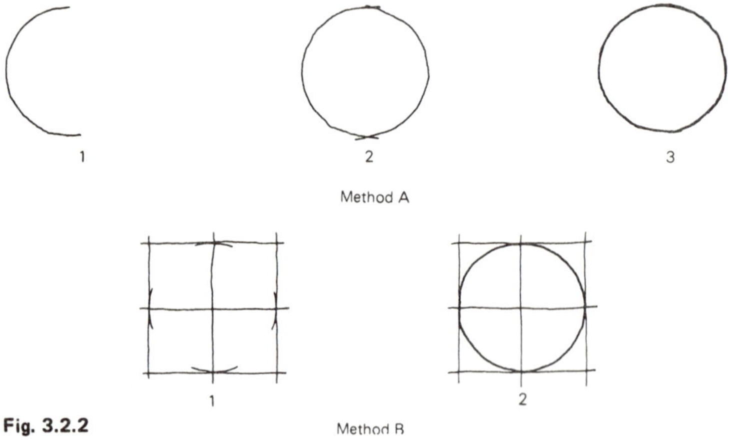

Fig. 3.2.2

Many vehicle components are cylindrical and when these are shown in pictorial form the circle becomes an ellipse. This ellipse always has a major axis at an angle of 90° to the cylinder centre line. The cylinder may be drawn in the stages shown in Fig. 3.2.3.

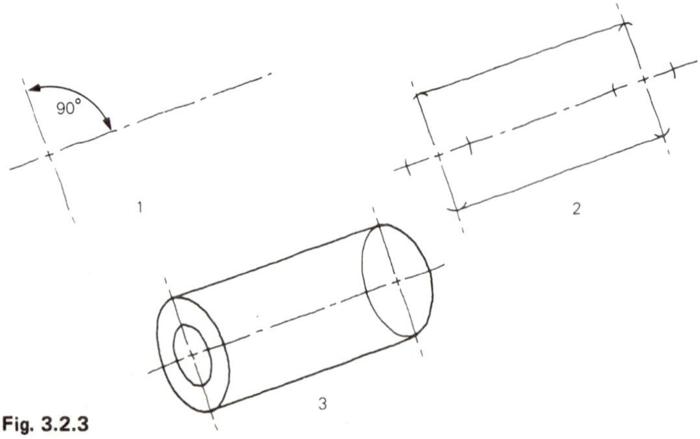

Fig. 3.2.3

When sketching automobile components it is advisable to:

(a) insert the centre line first,

(b) lightly sketch-in the complete object to obtain the correct proportion before lining-in,

(c) start at the centre and work outwards.

3.3 Line diagrams

A line diagram is a diagrammatic or schematic form of drawing which shows the main arrangement without unnecessary detail. Many uses of this method of drawing are found in the Technology subjects especially in the earlier years. Fig. 3.3.1a shows a typical example.

Electrical circuits are generally shown in line form; the cable shown by a straight line drawn either horizontally or vertically. Fig. 3.3.1b shows a circuit for a fuel gauge.

Line diagrams are used in conjunction with *block diagrams.* Fuel and lubrication systems are shown in this way. The component is shown by a block or box and the piping is represented by a line (Fig. 3.3.2).

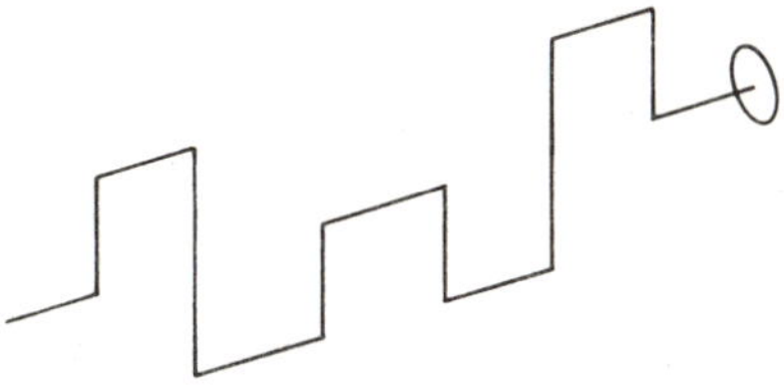
(a)

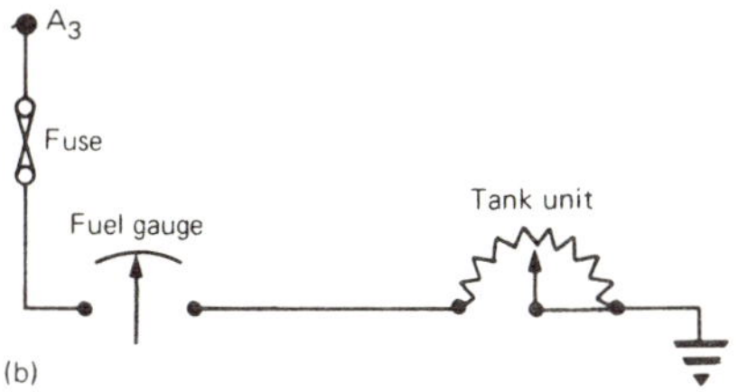

(b)

Fig. 3.3.1

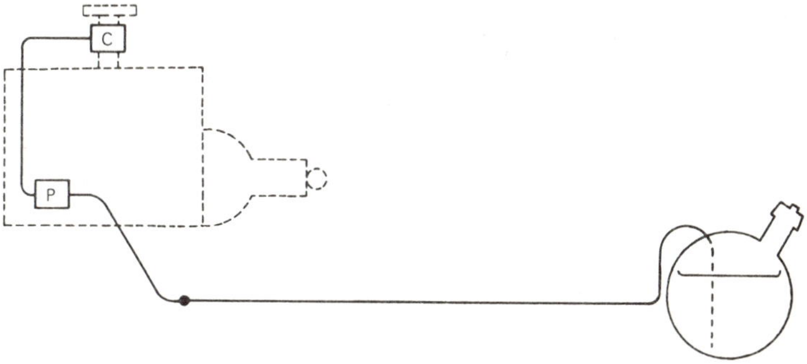

Fig. 3.3.2

3.4 Dimensioning

A dimension is provided on a drawing to give some other person information about the size or shape of an object. It is essential that this guidance can be easily understood, so rules are laid down in B.S. 308 for the manner in which the dimension is shown.

Consider the dimensions shown in Fig. 3.4.1.

The dimension line should be a thin, full line terminating with an arrow head not less than 3 mm long. Arrow heads should just touch the projection line. This is a thin full line starting just clear of the outline and extending beyond the arrow head.

Centre lines should not be used as dimension lines. The end elevation in Fig. 3.4.1 shows the dimension line moved around slightly.

If possible the dimension lines should be placed well away from the drawing with the largest dimension on the outside. The line may

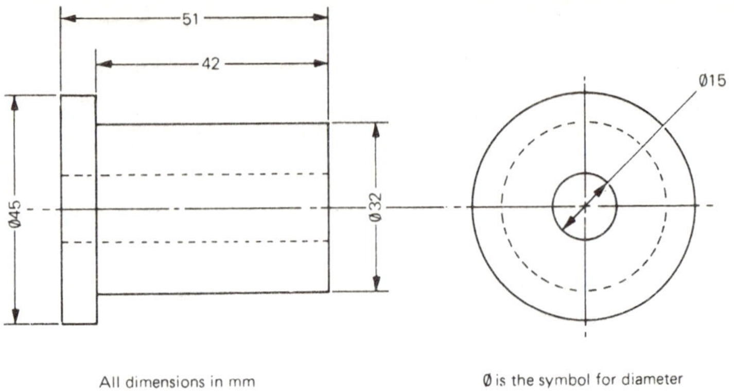

Fig. 3.4.1

be interrupted for the dimension or this may be placed above the line.

All dimensions should be arranged so that they can be read from either the bottom or the right-hand side of the drawing.

Fig. 3.4.2 shows a dimensioned drawing.

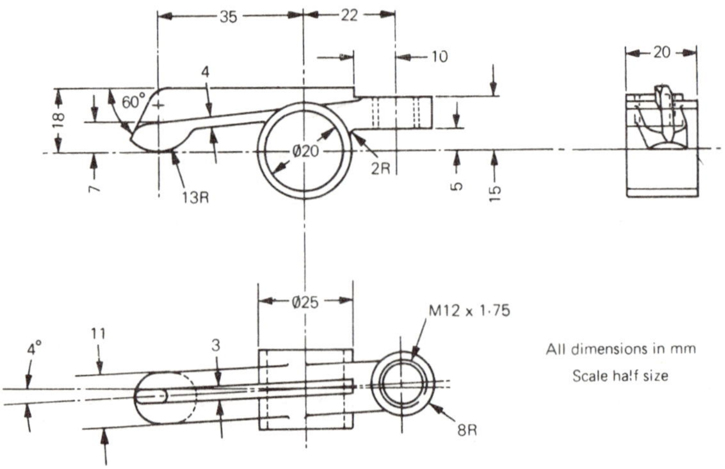

Fig. 3.4.2

3.5 Geometric construction

Constructions of the form shown in this chapter familiarise the student with his instruments and provide him with methods for overcoming many drawing and marking-out problems. Only the drawing aspect is considered in the following examples and since some of the methods are unsuitable for workshop application, it is suggested that the reader should apply his knowledge of marking-out tools to suggest an alternative workshop method.

Parallel lines

To draw a line parallel to a given line (Fig. 3.5.1).

(a) Set the compass to the distance required between the parallel lines. With the compass point on the line AB draw two arcs. Draw a straight line just touching the arcs. (b) Alternative method using set square and straight edge (Fig. 3.5.2). Hold the set-square against the

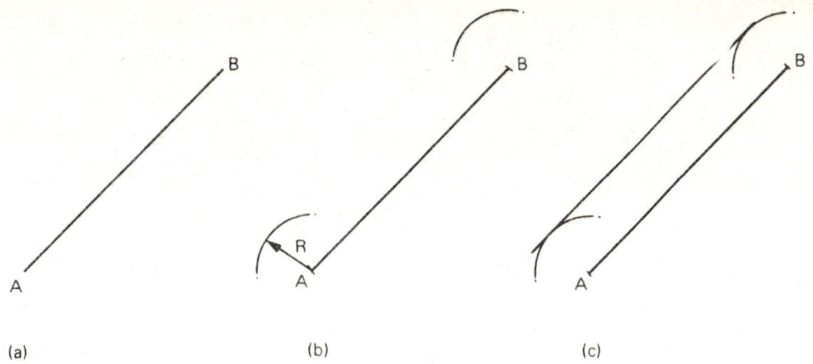

Fig. 3.5.1

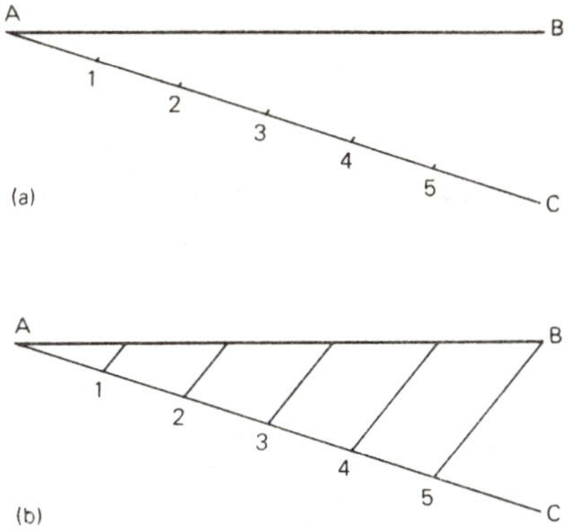

Fig. 3.5.3

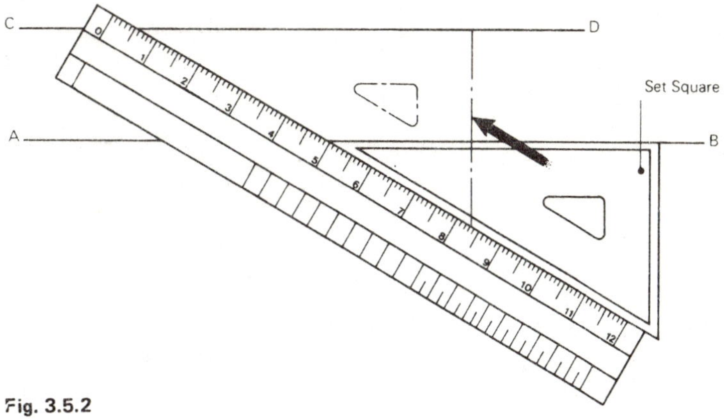

Fig. 3.5.2

straight edge with the edge of the square on the line. Without moving the straight edge slide the set square up the straight edge a given amount and draw the parallel line.

Division of a line

To divide a line into a given number of equal parts (Fig. 3.5.3).

Draw a construction line AC at any acute angle to the line AB. Using a rule or compass, step off the required number of divisions. With a set-square arranged as in Construction 1b, connect the last division with B and draw lines parallel to BC to give the required divisions along AB.

Perpendicular bisector

To bisect a given line by erecting a perpendicular (Fig. 3.5.4).

'To bisect' means to divide into two equal parts. With the compass point positioned at the spot where the perpendicular is to be erected, strike two arcs A and B. Adjust the compass to a radius of approximately AB and with the compass point positioned at the point where arc A crosses the line, strike arcs C and D. Repeat the operation with the compass point on arc B to give two further arcs E and F. Draw a straight line connecting the two intersection points of the arcs.

Bisecting an angle

To divide an angle into two equal angles (Fig. 3.5.5)..

Set the compass to a convenient radius and with the compass point positioned at the apex of the angle, draw an arc AB.

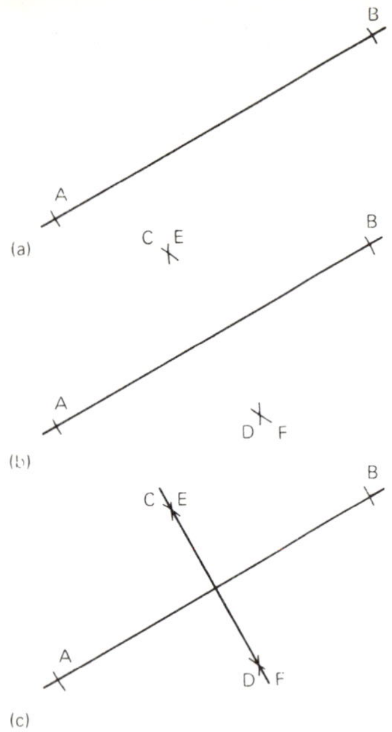

Fig. 3.5.4

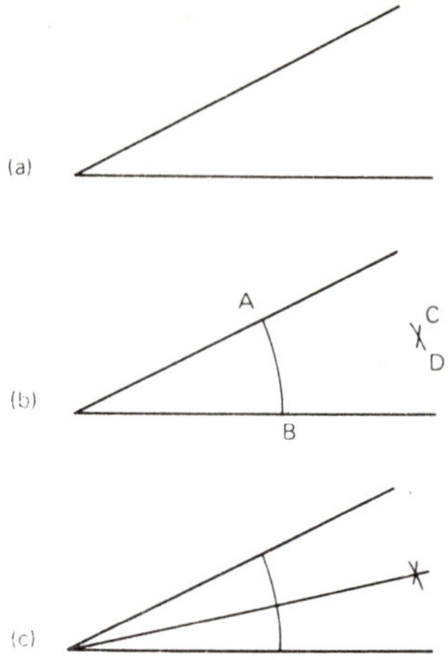

Fig. 3.5.5

Adjust the compass to a radius of approximately AB, position compass point at intersection A and draw arc C. Repeat with the compass at B to draw arc D. Connect intersection of arcs C and D with the apex of the angle.

Centre of arc

To find the centre of a given arc or circle (Fig. 3.5.6).

Draw two lines AB and CD of any convenient length. (These lines are called chords.) Place the compass at the point of intersection between the arc and chord and erect a perpendicular bisector on each line.

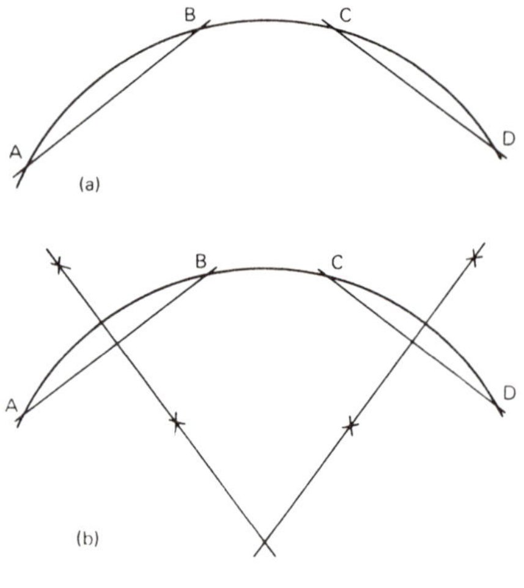

Fig. 3.5.6

The place where the two perpendiculars cross is the centre of the arc or circle.

Blending arc

To draw an arc of given radius to blend with two diverging lines (Fig. 3.5.7).

Using Construction 1 draw a line parallel to each diverging line at a distance equal to the radius of the required arc.

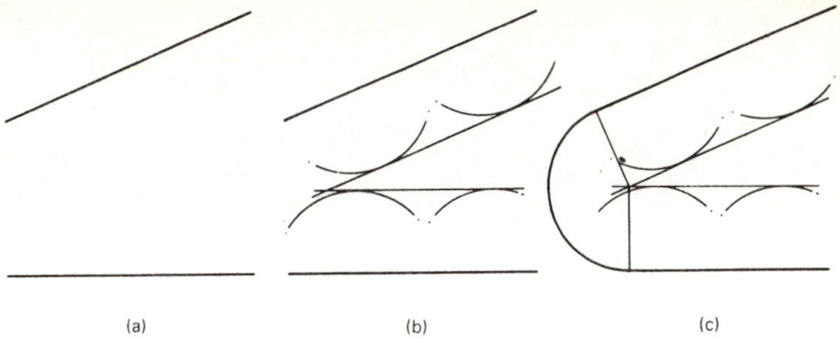

Fig. 3.5.7

Where the lines cross is the centre of the arc.

Right-angled triangle

To construct a right-angled triangle in a semi-circle (Fig. 3.5.8).

Draw a chord of any length in the semi-circle and draw another chord to form a triangle. The angle C is always a right angle.

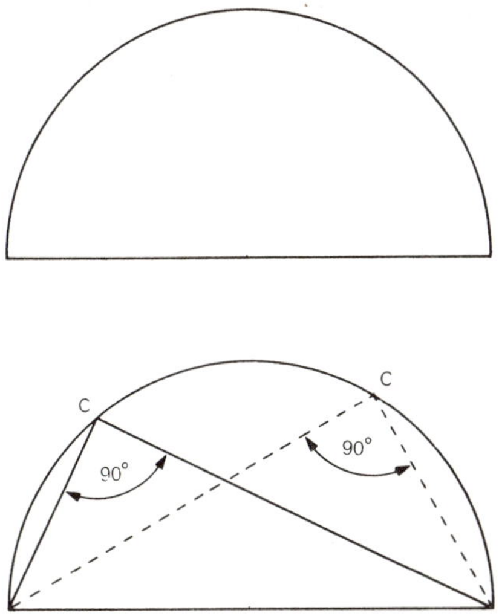

Fig. 3.5.8

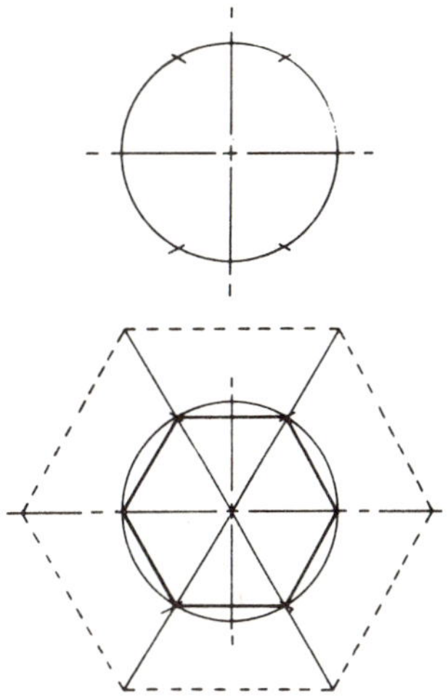

Fig. 3.5.9

Hexagon

To construct a regular hexagon (Fig. 3.5.9).

Draw a circle and without altering the compass setting, strike arcs to contact the circumference both sides of the centre line. Draw chords to join up the six points.

Lines can be projected outwards from the centre through each corner if a hexagon larger than the circle is required.

3.6 Screw threads

A screw thread is an application of the principle of the inclined plane or wedge. This can be easily demonstrated by wrapping a triangular piece of paper around a cylindrical object (Fig. 3.6.1). The helical form produced provides a compact form of wedge.

Conventional representation

B.S. 308 recommends that screw threads are drawn in the form shown in Fig. 3.6.2. The same Standard suggests the manner in which other items are represented. Fig. 3.6.3 shows the form used

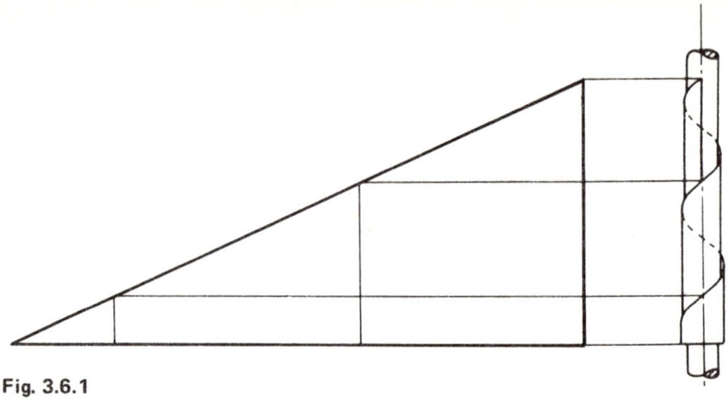

Fig. 3.6.1

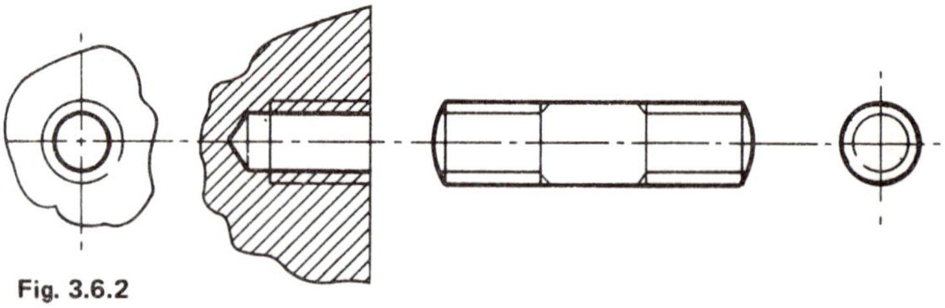

Fig. 3.6.2

a bearing and two mating gearwheels. The recommended convention saves time and enables the drawing to be easily understood.

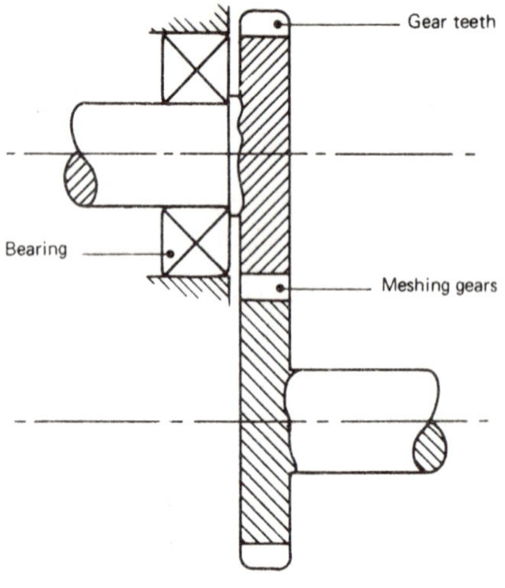

Fig. 3.6.3

Screw thread terms

Pitch — the distance between the crests of adjacent threads.
Lead — the amount the nut advances for one complete turn. In the case of the single start thread:

Pitch = lead

Core diameter — the diameter of the thread measured at the root. Fig. 3.6.4 shows that if an internal thread is to be formed in a piece of metal, the size of the hole must not be larger than the core

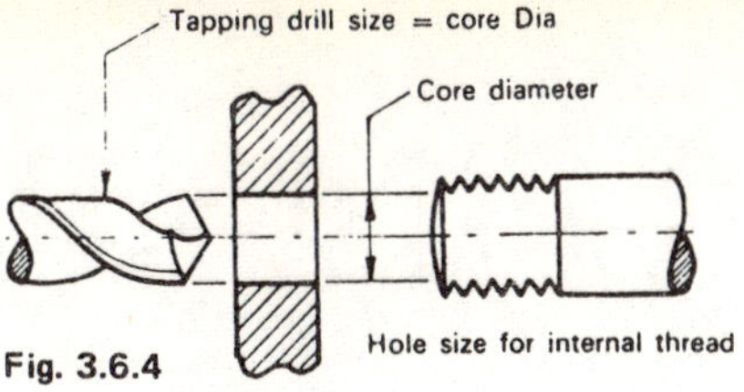

Fig. 3.6.4

Types of thread

diameter. Screw-thread tables give the size of the tapping drill required to suit a particular thread form. If a table is not available then the approximate size may be determined by either:

(i) finding the largest drill which just pass through a new nut of the appropriate size and thread form, or

(ii) measuring the core diameter of the thread cut on a taper tap.

Metric. International Standards Organisation (ISO) recommend the thread form shown as Fig. 3.6.5. The thread is made in either a course or fine pitch to suit the application. Information about the thread is given in the designation, e.g.

M8 x 1·25

M – ISO Metric thread
8 – diameter of bolt mm
1·25 – pitch mm

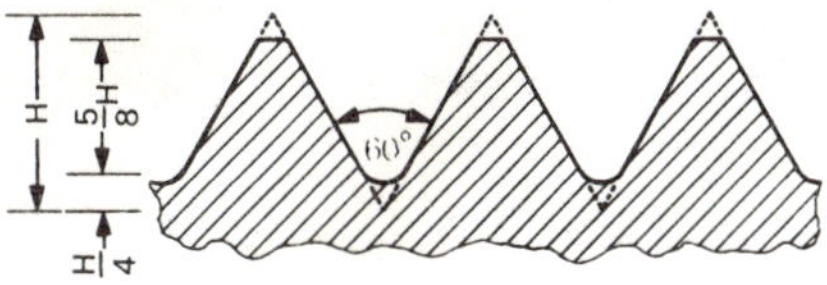

Fig. 3.6.5

The following table gives details of some screw thread sizes for general use on motor vehicles.

Bolt diameter mm	*Pitch* mm	
	Course	Fine
8	1·25	1
10	1·5	1·25
12	1·75	1·25
14	2	1·5
16	2	1·5

Unified threads are still in use on vehicles and Fig. 3.6.6a shows the profile of this type. The thread is made in a fine (UNF) or course (UNC) pitch; the UN is abbreviated from Unified National.

Other screw threads which have been commonly used in the past are shown in the table

Abbreviation	*Name*	*Thread angle*
BSW	British Standard Whitworth	55°
BSF	British Standard Fine	55°
BSP	British Standard Pipe	55°
BA	British Association	$47\frac{1}{2}$°

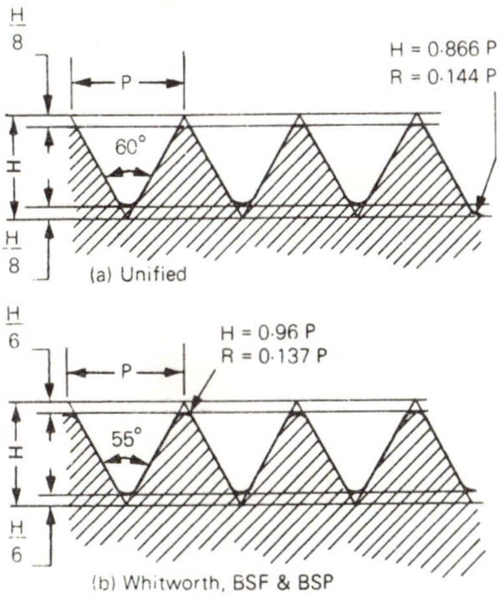

Fig. 3.6.6

3.7 Locking devices

Motor vehicles are subject to vibration and one effect of this is to cause securing devices such as nuts and bolts to work loose. The loss of a nut could cause either a serious accident or extensive component damage or both, so an effective locking device should be provided to reduce this risk.

Fig. 3.7.1 shows some of the main types.

a) *Spring washer.* Made of spring steel the washer digs into the nut to prevent the nut unscrewing. Various forms are in use:

(i) single coil

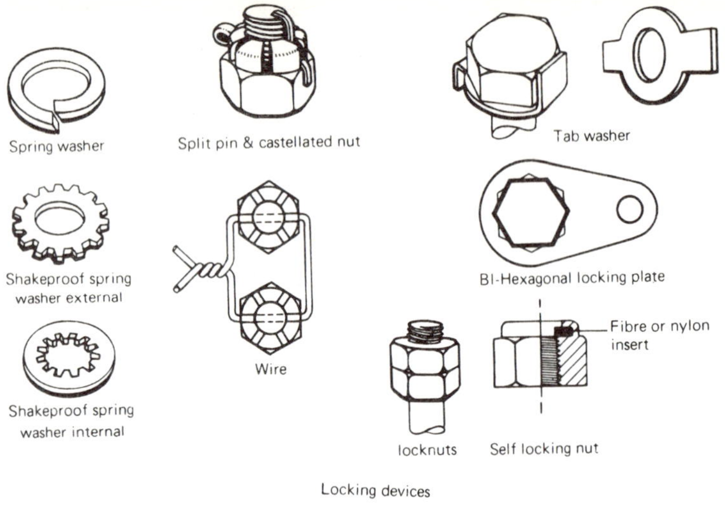

Fig. 3.7.1

(ii) double coil
(iii) shake-proof

b) *Split pin.* A mild steel pin fitted in a hole in a bolt engages with slots in a nut which is either:

(i) slotted – slots cut in a standard thickness nut
(ii) castellated – standard thickness nut + castellations.

Split pins, or cotter pins as they are sometimes called, should:

(i) be the correct diameter and length
(ii) only be used once
(iii) be fitted as shown in the diagram
(iv) not move in the nut when fitted.

c) *Wire.* Used to retain a nut to a stud and also prevents the stud from unscrewing.
d) *Tab washers.* Made of mild steel a tab washer has a tongue or tab to resist rotation and one or more external tabs which can be forced against the flat of the nut.
Two or more tab washers are sometimes joined together to form one large locking tab.

e) *Locking plate.* Generally a bi-hexagonal plate which fits over the nut.

f) *Lock nuts.* Two nuts screwed up against each other. The first nut is tightened and then the second nut is pulled up against the first to take up the slackness in the thread. This action means that the thread of the second nut resists the main load.

g) *Self-locking nuts.* Many different types are used. Normally they have some arrangement which increases the friction between the threads. The examples shown are:

(i) fibre or nylon insert
(ii) split

Special liquid compounds can be applied to the threads of a standard nut to provide a locking action.

Applications of these devices can be seen in F of M.V.T.

3.8 Sectional views

Technological work often deals with the detailed construction of components, so a drawing showing only the external view of a part has a limited use. Although the internal construction or hidden detail could be shown by a line of short dashes, the drawing could cause confusion. To overcome this drawback a sectioned view of the component is normally given. This shows the item either cut in half or cut through a plane which brings out some special feature of the design.

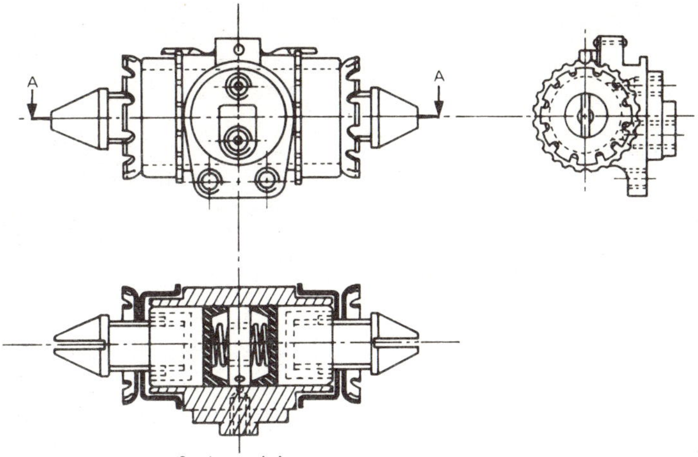

Fig. 3.8.1

Fig. 3.8.1 shows an external view, front and end elevation, of a brake cylinder. The plan shows the cylinder sectioned through the centre line. Not only will this drawing show the advantage of a sectional view but it will also indicate the main rules which apply to this method of representation.

(a) The sectioned drawing shows the part that remains after the portion nearest the reader is cut away.
(b) Material which is cut is indicated by section lining (cross hatching). These are thin lines, suitably spaced, drawn at an angle of 45° or some other angle to avoid the hatching running parallel with the outline. Adjacent sectioned portions should either be lined in different directions or at contrasting pitches.
(c) The cutting plane, shown by a long chain line, is lettered and arrowed to show the direction in which the section is viewed.
(d) Components which are symmetrical about a centre line are normally drawn in half section (Fig. 3.8.2).
(e) When a cutting plane passes longitudinally through ribs, shafts, bolts, nuts, rods, rivets, keys, balls, rollers and split pins, the items mentioned are not sectioned.

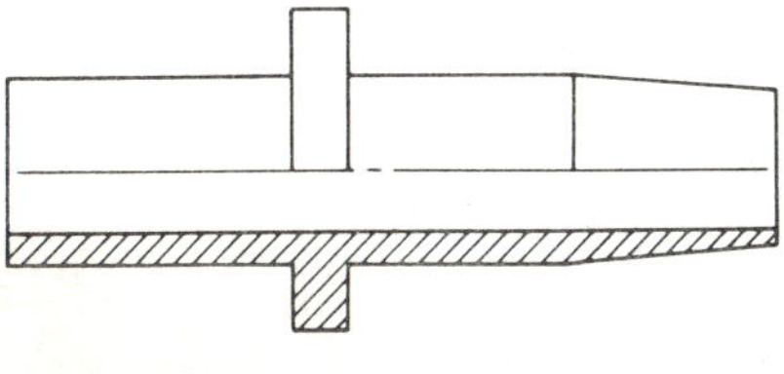

Fig. 3.8.2

3.9 Limits and tolerances

Consider the measurement of a stud which is stated to have a length of 50 mm.

Although great care might have been exercised in its production it it is most unlikely that the actual length of the stud is precisely 50 mm. A rule may indicate a size of 50 mm, but when the specimen is checked with a micrometer, it will most probably show that the length is not 50 mm. In a similar manner, if a micrometer was used to gauge the length during manufacture and more precise equipment is then used to check the length, it will once again show up the inaccuracy. This example shows that very few components could be made to a definite dimension and indicates how the accuracy of a product is affected by the type of measuring instrument used.

The designer must decide on the accuracy required. His decision will be governed by the effect of any inaccuracy of a part on other mating parts and on the amount of interchangeability required. He will be aware that a high degree of accuracy causes high production costs and this means that the standard suggested should not be higher than necessary.

Limits

After fixing the nominal dimension, which in the case of the stud previously mentioned was 50 mm the designer decides on the limits of size. He states the largest and smallest size which are acceptable and these limits could be shown as:

$$50 \pm 0{\cdot}05 \quad \text{or} \quad \begin{matrix} 50{\cdot}05 \\ 49{\cdot}95 \end{matrix} \quad \text{or} \quad \begin{matrix} 50{\cdot}05 \\ -0{\cdot}1 \end{matrix}$$

In this example the size of the part must fall between the limits 49·95 and 50·05. The largest and smallest size are called the upper and lower limit respectively. Limits do not always fall on each side of the nominal dimension; for example:

$$50^{-0{\cdot}05}_{-0{\cdot}1} \quad \text{or} \quad \begin{matrix} 49{\cdot}95 \\ 49{\cdot}9 \end{matrix} \quad \text{or} \quad \begin{matrix} 49{\cdot}95 \\ -0{\cdot}05 \end{matrix}$$

In all cases the largest dimension is placed at the top.

Tolerance

'Tolerance' is the acceptable size range into which the component must fall; in other words, the range between upper and lower limits.

An item dimensioned $50^{+0{\cdot}12}_{-0{\cdot}05}$ has a tolerance of:

$$\text{Upper limit} - \text{lower limit} = 50{\cdot}12 - 49{\cdot}95 = 0{\cdot}17$$

The tolerance indicates the degree of accuracy required.

Examples

1. What is the maximum size, minimum size and tolerance allowed for the following:

	Max. size	*Min. size*	*Tolerance*
(a) 11·5 ± 0·005	11·505	11·495	0·01
(b) $13{\cdot}2^{+0{\cdot}013}_{-0{\cdot}004}$	13·213	13·196	0·017

(c) $12{\cdot}5^{+0{\cdot}007}_{+0{\cdot}003}$	12·507	12·503	0·004
(d) $16{\cdot}3^{\;0}_{-0{\cdot}008}$	16·3	16·292	0·008
(e) $11{\cdot}7_{-0{\cdot}005}$	11·7	11·695	0·005

2. What is the maximum angle, minimum angle and tolerance allowed for the following:

	Max. angle	*Min. angle*	*Tolerance*
(a) $90° \pm 5°$	95°	85°	10°
(b) $\begin{matrix}45°\\44°\end{matrix}$	45°	44°	1°
(c) $\begin{matrix}22°\ 0'\\-0°\ 30'\end{matrix}$	22°	21° 30′	0° 30′

Types of fit

The relative size of two mating parts governs the fit. Fig. 3.9.1 shows two basic types of fit.

Clearance fit

The shaft is always smaller than the hole; the size difference between the shaft and hole is termed the *clearance*. Tolerance on the shaft and hole can cause the clearance to vary by a good deal. If the tolerance on the parts is large, then *selective assembly* is necessary if a specific clearance is to be obtained. This method of assembly entails the mating-up of a large shaft with a large hole, and so on.

Motor vehicle components such as pistons and crankshafts are examples of a clearance fit.

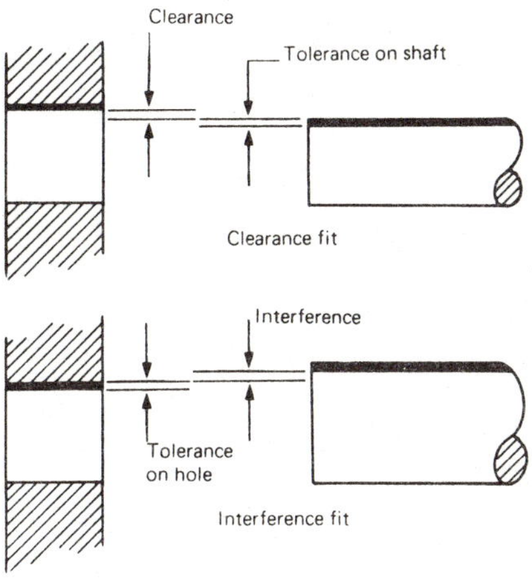

Figure 3.9.1

Interference fit

The shaft is larger than the hole; the difference in size is termed the *interference.*

The shaft is inserted in the hole by:

(a) utilising force given by a press or hammer,
(b) heating the female member,
(c) cooling the male member.

This type of fit is often used to retain a part in position. Items such as a cylinder liner and flywheel ring gear are examples of an interference fit.

Exercises 3.10

1. A drawing which shows three separate views; front elevation, end elevation and plan, would be called:

(a) oblique projection
(b) isometric projection
(c) pictorial projection
(d) orthographic projection

2. The term '1st angle' is one form of:

(a) oblique projection
(b) isometric projection
(c) pictorial projection
(d) orthographic projection

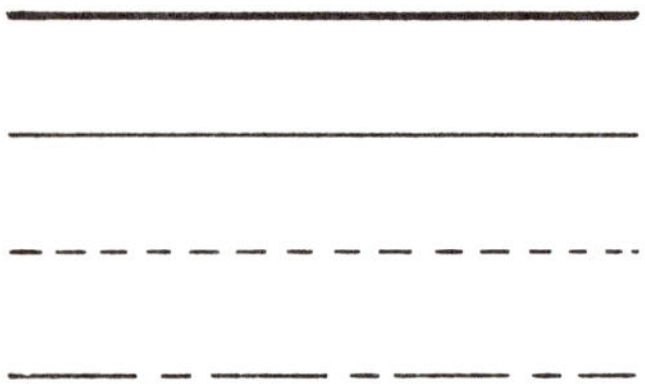

Fig. 3.10.1

3. The view obtained 'from looking down on an object' is positioned on the drawing paper under the front elevation. This projection is called:
(a) 1st angle (c) isometric
(b) 3rd angle (d) oblique

4. A centre line on a drawing is represented in Fig. 3.10.1 by the number:
(a) 1 (c) 3
(b) 2 (d) 4

5. A visible outline on a drawing is represented in Fig. 3.10.1 by the number:
(a) 1 (c) 3
(b) 2 (d) 4

6. Hidden detail in a drawing is represented in Fig. 3.10.1 by the number:
(a) 1 (c) 3
(b) 2 (d) 4

7. It is recommended that dimensions should be inserted in a drawing so that they can be read from either the:
(a) bottom or left hand side (c) top or right hand side
(b) bottom or right hand side (d) top or right hand side

Questions 8 — 11 refer to Fig. 3.4.1

8. The component has an overall length of:
(a) 32 mm (c) 45 mm
(b) 42 mm (d) 51 mm

9. The dimensions of the hole in the component indicate that the hole has a diameter of:
(a) 15 mm and a length of 51 mm
(b) 15 mm and a length of 42 mm
(c) 32 mm and a length of 42 mm
(d) 45 mm and a length of 51 mm

10. The drawing uses a projection which is called:
(a) 1st angle (c) 3rd angle
(b) 2nd angle (d) 4th angle

11. The drawing shows that the component has a boss of diameter 45 mm. What is the thickness of this boss?

(a) 9 mm (c) 42 mm
(b) 32 mm (d) 51 mm

Questions 12—18 refer to Fig. 3.4.2

12. What is the distance in mm between the centre line of the bush and the centre line of the threaded portion?
(a) 5 (c) 20
(b) 10 (d) 22

13. A screw engaged in the female thread of the component is first moved to a position where all of the threads are in contact and then

it is completely unscrewed. The minimum number of turns needed to remove the screw is:

(a) 1·25 (c) 8
(b) 5·72 (d) 8·57

14. What is the length in mm of the hole of diameter 20 mm?

(a) 11 (c) 22
(b) 20 (d) 25

15. What is the angle in degrees between the centre line of the large hole and the centre line drawn from the thread through the 3 mm thick web?

(a) 3 (c) 60
(b) 11 (d) 86

16. The largest curved surface has a radius of:

(a) 8 mm (c) 12·5 mm
(b) 10 mm (d) 13 mm

17. The pitch of a screw thread is the:

(a) diameter of the core of the screw
(b) difference between core and crest diameters
(c) distance between the crests of adjacent threads
(d) length of the threaded part of a bolt

18. When a thread has to be tapped in an object, the size of the 'tapping drill' should:

(a) be slightly larger than the crest diameter of the bolt
(b) not be larger than the core diameter of the thread
(c) not be smaller than the crest diameter of the thread
(d) be equal to the pitch of the thread

19. What is meant by the term 'limit' as applied to dimensions?

(a) The difference between smallest and largest acceptable size
(b) The difference between maximum and nominal sizes
(c) The variation from the nominal size
(d) The maximum and minimum size of a component

20. What is meant by the term 'tolerance' as applied to dimensions?

(a) The range between upper and lower limits
(b) The maximum and minimum size of a component
(c) The difference between maximum and nominal sizes
(d) The difference in size between a shaft and hole

21. The maximum size and tolerance for a component dimensioned 13·5 ± 0·003 are

(a) 13·8 and 0·006 (c) 13·503 and 0·006
(b) 13·8 and 0·003 (d) 13·203 and 0·006

22. The maximum and minimum sizes for a component dimensioned $22{\cdot}1\,^{+0{\cdot}006}_{-0{\cdot}001}$ are:

(a) max 22·106 min 22·009 (c) max 22·6 min 21·9
(b) max 22·106 min 22·099 (d) max 22·7 min 22

23. The maximum size, minimum size and tolerance,for a component dimensioned $44{\cdot}7\,^{+0{\cdot}009}_{+0{\cdot}005}$ is:

(a) max 44·709	min 44·695	tolerance 0·014
(b) max 44·709	min 44·705	tolerance 0·004
(c) max 44·79	min 44·75	tolerance 0·04
(d) max 45·6	min 44·2	tolerance 0·4

24. The maximum size, minimum size and tolerance for a component dimensioned $13{\cdot}4\,^{-0}_{-0{\cdot}006}$ is:

(a) max 13:394	min 13·388	tolerance 0·006
(b) max 13·4	min 12·8	tolerance 0·6
(c) max 13·4	min 13·394	tolerance 0·006
(d) max 13·406	min 13·394	tolerance 0·012

25. What is meant by the term 'clearance' as applied to a shaft and bearing? The bearing is:
(a) larger than the shaft and the space between is the clearance
(b) smaller than the shaft and the space between is the clearance
(c) made the same size as the shaft and clearance is the wear
(d) made the same size as the shaft and clearance is the tolerance

26. What is meant by the term 'interference' as applied to a shaft and hole? The hole is:
(a) larger than the shaft and the difference in size is the interference
(b) smaller than the shaft and the difference in size is the interference
(c) made the same size as the shaft and interference is the wear
(d) made the same size as the shaft and interference is the tolerance

4 APPENDIX

4.1 Multi-choice questions

This type of question is widely used in Craft examinations so a few words of advice may help the reader to improve his technique of answering this type of question:

1. *Understand the instructions.* Examination papers contain precise instructions about such things as where and how the answers are presented together with other important details about the examination.

These instructions must be read and understood. If you are worried about any detail, the invigilator should be consulted.

Modern computors are used to 'mark' the papers and credit for correct answers may not be given if the answers are not presented in the manner specified.

2. *Read the question carefully.* Many examination candidates think that this type of question is very easy. Most probably they arrive at this conclusion due to the ease and speed in which they can present their answers, but results show that unless care is exercised, the candidates' result is far from satisfactory. The multi-choice paper demands extra special attention since if some small point in the question is overlooked then an incorrect answer is selected.

Before looking at the possible answers the question should be understood even if it means reading it over again.

3. *Selecting the answer.* When the question becomes clear the candidate will formulate the answer in his mind. At this stage he should examine each option given and then decide which one has a wording similar to that required. Alternatively the selection may be decided by applying the true/false test. This involves the reading of *each* option and mentally deciding if it is true or false. In this manner the possible answers can be narrowed down and where doubt exists the chance of selecting the correct answer is improved. This is quite different to guessing – the candidates knowledge and common sense are used to eliminate the incorrect statements.

Questions involving calculations should be treated in a similar way – the answer should be worked out in some clear space on the paper and having obtained the calculated answer the option nearest to this should be selected. If printed values do not agree with your result then checking is necessary. It is not intended from this that checking should not be performed in other cases. Remember that the other options are normally obtained by including the common errors made by candidates.

Multiple choice questions should only have one correct answer; the three other options are called distractors and their purpose is to present to the 'unsure' candidate an attractive possible answer.

4.2 Useful conversions

Imperial to SI

Most of the following conversions have been approximated to give values suitable for general use. If greater accuracy is required the internationally agreed equivalent should be used.

Length

1 yard is exactly 0·9144 metre.

1 inch (in) ≙ 25·4 mm — 1 foot (ft) ≙ 304·8 m
0·001 in ≙ 0·025 mm — mile ≙ 1·6 km

Area

1 sq inch (in^2) ≙ 6·45 cm^2 — 1 sq foot (ft^2) ≙ 929 cm^2

Volume

1 cubic inch (in^3) ≙ 16·4 cm^3 — 1 cubic ft (ft^3) ≙ 0·028 m^3

Capacity

1 pint (pt) ≙ 0·568 litre — 1 gallon (gal) ≙ 4·5 litre

Mass

1 pound (lb) is exactly 0·453 592 37 kg

1 ounce (oz) ≙ 28·35 g — 1 ton ≙ 1016 kg = 1·016 tonne

Force

1 pound force (lbf) ≙ 4·45 N — 1 ton ≙ 10 kN

Torque

1 pound foot (lbf ft) ≙ 1·4 Nm

Pressure and stress

1 pound/sq inch (lbf/in^2) ≙ 7 kN/m^2 = 7 kPa = 0·07 bar = 70 m bar
1 ton/sq inch (ton/in^2) ≙ 15 MN/m^2
1 atmosphere (atm) ≙ 100 kN/m^2 = 100 kPa = 1 bar
1 inch mercury (in Hg) ≙ 3·4 kN/m^2 = 3·4 kPa = 34 mbar

Work and energy

1 foot pound (ft lbf) ≙ 1·4 J
1 British thermal unit (Btu) ≙ 1055 J = 1·055 kJ
1 Centigrade heat unit (Chu) ≙ 1900 J = 1·9 kJ

Power

1 horsepower (hp) ≙ 746 W

Velocity

1 foot/second (ft/s) ≙ 0·3 m/s
1 mile/h (mile/h) ≙ 1·6 km/h

Acceleration

1 foot/second^2 (ft/s^2)	$\simeq$ 0·3 m/s^2

Consumption

1 mile/gallon (mile/gal)	$\simeq$ 0·34 km/litre OR 290 litres/100 km

SQUARE ROOTS

	0	1	2	3	4	5	6	7	8	9	Mean Differences 1	2	3	4	5	6	7	8	9
1·0	1·000	1·005	1·010	1·015	1·020	1·025	1·030	1·034	1·039	1·044	0	1	1	2	2	3	3	4	4
1·1	1·049	1·054	1·058	1·063	1·068	1·072	1·077	1·082	1·086	1·091	0	1	1	2	2	3	3	4	4
1·2	1·095	1·100	1·105	1·109	1·114	1·118	1·122	1·127	1·131	1·136	0	1	1	2	2	3	3	4	4
1·3	1·140	1·145	1·149	1·153	1·158	1·162	1·166	1·170	1·175	1·179	0	1	1	2	2	3	3	3	4
1·4	1·183	1·187	1·192	1·196	1·200	1·204	1·208	1·212	1·217	1·221	0	1	1	2	2	2	3	3	4
1·5	1·225	1·229	1·233	1·237	1·241	1·245	1·249	1·253	1·257	1·261	0	1	1	2	2	2	3	3	4
1·6	1·265	1·269	1·273	1·277	1·281	1·285	1·288	1·292	1·296	1·300	0	1	1	2	2	2	3	3	3
1·7	1·304	1·308	1·311	1·315	1·319	1·323	1·327	1·330	1·334	1·338	0	1	1	2	2	2	3	3	3
1·8	1·342	1·345	1·349	1·353	1·356	1·360	1·364	1·367	1·371	1·375	0	1	1	1	2	2	3	3	3
1·9	1·378	1·382	1·386	1·389	1·393	1·396	1·400	1·404	1·407	1·411	0	1	1	1	2	2	3	3	3
2·0	1·414	1·418	1·421	1·425	1·428	1·432	1·435	1·439	1·442	1·446	0	1	1	1	2	2	2	3	3
2·1	1·449	1·453	1·456	1·459	1·463	1·466	1·470	1·473	1·476	1·480	0	1	1	1	2	2	2	3	3
2·2	1·483	1·487	1·490	1·493	1·497	1·500	1·503	1·507	1·510	1·513	0	1	1	1	2	2	2	3	3
2·3	1·517	1·520	1·523	1·526	1·530	1·533	1·536	1·539	1·543	1·546	0	1	1	1	2	2	2	3	3
2·4	1·549	1·552	1·556	1·559	1·562	1·565	1·568	1·572	1·575	1·578	0	1	1	1	2	2	2	3	3
2·5	1·581	1·584	1·587	1·591	1·594	1·597	1·600	1·603	1·606	1·609	0	1	1	1	2	2	2	3	3
2·6	1·612	1·616	1·619	1·622	1·625	1·628	1·631	1·634	1·637	1·640	0	1	1	1	2	2	2	2	3
2·7	1·643	1·646	1·649	1·652	1·655	1·658	1·661	1·664	1·667	1·670	0	1	1	1	2	2	2	2	3
2·8	1·673	1·676	1·679	1·682	1·685	1·688	1·691	1·694	1·697	1·700	0	1	1	1	1	2	2	2	3
2·9	1·703	1·706	1·709	1·712	1·715	1·718	1·720	1·723	1·726	1·729	0	1	1	1	1	2	2	2	3
3·0	1·732	1·735	1·738	1·741	1·744	1·746	1·749	1·752	1·755	1·758	0	1	1	1	1	2	2	2	3
3·1	1·761	1·764	1·766	1·769	1·772	1·775	1·778	1·780	1·783	1·786	0	1	1	1	1	2	2	2	3
3·2	1·789	1·792	1·794	1·797	1·800	1·803	1·806	1·808	1·811	1·814	0	1	1	1	1	2	2	2	2
3·3	1·817	1·819	1·822	1·825	1·828	1·830	1·833	1·836	1·838	1·841	0	1	1	1	1	2	2	2	2
3·4	1·844	1·847	1·849	1·852	1·855	1·857	1·860	1·863	1·865	1·868	0	1	1	1	1	2	2	2	2
3·5	1·871	1·873	1·876	1·879	1·881	1·884	1·887	1·889	1·892	1·895	0	1	1	1	1	2	2	2	2
3·6	1·897	1·900	1·903	1·905	1·908	1·910	1·913	1·916	1·918	1·921	0	1	1	1	1	2	2	2	2
3·7	1·924	1·926	1·929	1·931	1·934	1·936	1·939	1·942	1·944	1·947	0	1	1	1	1	2	2	2	2
3·8	1·949	1·952	1·954	1·957	1·960	1·962	1·965	1·967	1·970	1·972	0	1	1	1	1	2	2	2	2
3·9	1·975	1·977	1·980	1·982	1·985	1·987	1·990	1·995	1·995	1·997	0	1	1	1	1	2	2	2	2
4·0	2·000	2·002	2·005	2·007	2·010	2·012	2·015	2·017	2·020	2·022	0	0	1	1	1	1	2	2	2
4·1	2·025	2·027	2·030	2·032	2·035	2·037	2·040	2·042	2·045	2·047	0	0	1	1	1	1	2	2	2
4·2	2·049	2·052	2·054	2·057	2·059	2·062	2·064	2·066	2·069	2·071	0	0	1	1	1	1	2	2	2
4·3	2·074	2·076	2·078	2·081	2·083	2·086	2·088	2·090	2·093	2·095	0	0	1	1	1	1	2	2	2
4·4	2·098	2·100	2·102	2·105	2·107	2·110	2·112	2·114	2·117	2·119	0	0	1	1	1	1	2	2	2
4·5	2·121	2·124	2·126	2·128	2·131	2·133	2·135	2·138	2·140	2·142	0	0	1	1	1	1	2	2	2
4·6	2·145	2·147	2·149	2·152	2·154	2·156	2·159	2·161	2·163	2·166	0	0	1	1	1	1	2	2	2
4·7	2·168	2·170	2·173	2·175	2·177	2·179	2·182	2·184	2·186	2·189	0	0	1	1	1	1	2	2	2
4·8	2·191	2·193	2·195	2·198	2·200	2·202	2·205	2·207	2·209	2·211	0	0	1	1	1	1	2	2	2
4·9	2·214	2·216	2·218	2·220	2·225	2·227	2·229	2·232	2·232	2·234	0	0	1	1	1	1	2	2	2
5·0	2.236	2·238	2·241	2·243	2·245	2·247	2·249	2·252	2·254	2·256	0	0	1	1	1	1	2	2	2
5·1	2·258	2·261	2·263	2·265	2·267	2·269	2·272	2·274	2·276	2·278	0	0	1	1	1	1	2	2	2
5·2	2·280	2·283	2·285	2·287	2·289	2·291	2·293	2·296	2·298	2·300	0	0	1	1	1	1	2	2	2
5·3	2·302	2·304	2·307	2·309	2·311	2·313	2·315	2·317	2·319	2·322	0	0	1	1	1	1	2	2	2
5·4	2·324	2·326	2·328	2·330	2·332	2·335	2·337	2·339	2·341	2·343	0	0	1	1	1	1	2	2	2

SQUARE ROOTS

	0	1	2	3	4	5	6	7	8	9	Mean Differences 1	2	3	4	5	6	7	8	9
5·5	2·345	2·347	2·349	2:352	2·354	2·356	2·358	2·360	2·362	2·364	0	0	1	1	1	1	1	2	2
5·6	2·366	2·369	2·371	2·373	2·375	2·377	2·379	2·381	2·383	2·385	0	0	1	1	1	1	1	2	2
5·7	2·387	2·390	2·392	2·394	2·396	2·398	2·400	2·402	2·404	2·406	0	0	1	1	1	1	1	2	2
5·8	2·408	2·410	2·412	2·415	2·417	2·419	2·421	2·423	2·425	2·427	0	0	1	1	1	1	1	2	2
5·9	2·429	2·431	2·433	2·435	2·437	2·439	2·441	2·443	2·445	2·447	0	0	1	1	1	1	1	2	2
6·0	2·449	2·452	2·454	2·456	2·458	2·460	2·462	2·464	2·466	2·468	0	0	1	1	1	1	1	2	2
6·1	2·470	2·472	2·474	2·476	2·478	2·480	2·482	2·484	2·486	2·488	0	0	1	1	1	1	1	2	2
6·2	2·490	2·492	2·494	2·496	2·498	2·500	2·502	2·504	2·506	2·508	0	0	1	1	1	1	1	2	2
6·3	2·510	2·512	2·514	2·516	2·518	2·520	2·522	2·524	2·526	2·528	0	0	1	1	1	1	1	2	2
6·4	2·530	2·532	2·534	2·536	2·538	2·540	2·542	2·544	2·546	2·548	0	0	1	1	1	1	1	2	2
6·5	2·550	2·551	2·553	2·555	2·557	2·559	2·561	2·563	2·565	2·567	0	0	1	1	1	1	1	2	2
6·6	2·569	2·571	2·573	2·575	2·577	2·579	2·581	2·583	2·585	2·587	0	0	1	1	1	1	1	2	2
6·7	2·588	2·590	2·592	2·594	2·596	2·598	2·600	2·602	2·604	2·606	0	0	1	1	1	1	1	2	2
6·8	2·608	2·610	2·612	2·613	2·615	2·617	2·619	2·621	2·623	2·625	0	0	1	1	1	1	1	2	2
6·9	2·627	2·629	2·631	2·632	2·634	2·636	2·638	2·640	2·642	2·644	0	0	1	1	1	1	1	2	2
7·0	2·646	2·648	2·650	2·651	2·653	2·655	2·657	2·659	2·661	2·663	0	0	1	1	1	1	1	2	2
7·1	2·665	2·666	2·668	2·670	2·672	2·674	2·676	2·678	2·680	2·681	0	0	1	1	1	1	1	1	2
7·2	2·683	2·685	2·687	2·698	2·691	2·693	2·694	2·696	2·698	2·700	0	0	1	1	1	1	1	1	2
7·3	2·702	2·704	2·706	2·707	2·709	2·711	2·713	2·715	2·717	2·718	0	0	1	1	1	1	1	1	2
7·4	2·720	2·722	2·724	2·726	2·728	2·729	2·731	2·733	2·735	2·737	0	0	1	1	1	1	1	1	2
7·5	2·739	2·740	2·742	2·744	2·746	2·748	2·750	2·751	2·753	2·755	0	0	1	1	1	1	1	1	2
7·6	2·757	2·759	2·760	2·762	2·764	2·766	2·768	2·769	2·771	2·773	0	0	1	1	1	1	1	1	2
7·7	2·775	2·777	2·778	2·780	2·782	2·784	2·786	2·787	2·789	2·791	0	0	1	1	1	1	1	1	2
7·8	2·793	2·795	2·796	2·798	2·800	2·802	2·804	2·805	2·807	2·809	0	0	1	1	1	1	1	1	2
7·9	2·811	2·812	2·814	2·816	2·818	2·820	2·821	2·823	2·825	2·827	0	0	1	1	1	1	1	1	2
8·0	2·828	2·830	2·832	2·834	2·835	2·837	2·839	2·841	2·843	2·844	0	0	1	1	1	1	1	1	2
8·1	2·846	2·848	2·850	2·851	2·853	2·855	2·857	2·858	2·860	2·862	0	0	1	1	1	1	1	1	2
8·2	2·864	2·865	2·867	2·869	2·871	2·872	2·874	2·876	2·877	2·879	0	0	1	1	1	1	1	1	2
8·3	2·881	2·883	2·884	2·886	2·888	2·890	2·891	2·893	2·895	2·897	0	0	1	1	1	1	1	1	2
8·4	2·898	2·900	2·902	2·903	2·905	2·907	2·909	2·910	2·912	2·914	0	0	1	1	1	1	1	1	2
8·5	2·915	2·917	2·919	2·921	2·922	2·924	2·926	2·927	2·929	2·931	0	0	1	1	1	1	1	1	2
8·6	2·933	2·934	2·936	2·938	2·939	2·941	2·943	2·944	2·946	2·948	0	0	1	1	1	1	1	1	2
8·7	2·950	2·951	2·953	2·955	2·956	2·958	2·960	2·961	2·963	2·965	0	0	1	1	1	1	1	1	2
8·8	2·966	2·968	2·970	2·972	2·973	2·975	2·977	2·978	2·980	2·982	0	0	1	1	1	1	1	1	2
8·9	2·983	2·985	2·987	2·988	2·990	2·992	2·993	2·995	2·997	2·998	0	0	1	1	1	1	1	1	2
9·0	3·000	3·002	3·003	3·005	3·007	3·008	3·010	3·012	3·013	3·015	0	0	0	1	1	1	1	1	1
9·1	3·017	3·018	3·020	3·022	3·023	3·025	3·027	3:028	3·030	3·032	0	0	0	1	1	1	1	1	1
9·2	3·033	3·035	3·036	3·038	3·040	3·041	3·043	0·045	3·046	3·048	0	0	0	1	1	1	1	1	1
9·3	3·050	3·051	3·053	3·055	3·056	3·058	3·059	3·061	3·063	3·064	0	0	0	1	1	1	1	1	1
9·4	3·066	3·068	3·069	3·071	3·072	3·074	3·076	3·077	3·079	3·081	0	0	0	1	1	1	1	1	1
9·5	3·082	3·084	3·085	3·087	3·089	3·090	3·092	3·094	3·095	3·097	0	0	0	1	1	1	1	1	1
9·6	3·098	3·100	3·102	3·103	3·105	3·106	3·108	3·110	3·111	3·113	0	0	0	1	1	1	1	1	1
9·7	3·114	3·116	3·118	3·119	3·121	3·122	3·124	3·126	3·127	3·129	0	0	0	1	1	1	1	1	1
9·8	3·130	3·132	3·134	3·135	3·137	3·138	3·140	3·142	3·145	3·145	0	0	0	1	1	1	1	1	1
9·9	3·146	3·148	3·150	3·151	3·153	3·154	3·156	3·158	3·159	3·161	0	0	0	1	1	1	1	1	1

SQUARE ROOTS

	0	1	2	3	4	5	6	7	8	9	Mean Differences								
											1	2	3	4	5	6	7	8	9
10	3·162	3·178	3·194	3·209	3·225	3·240	3·256	3·271	3·286	3·302	2	3	5	6	8	9	11	12	14
11	3·317	3·332	3·347	3·362	3·376	3·391	3·406	3·421	3·435	3·450	1	3	4	6	7	9	10	12	13
12	3·464	3·479	3·493	3·507	3·521	3·536	3·550	3·564	3·578	3·592	1	3	4	6	7	8	10	11	13
13	3·606	3·619	3·633	3·647	3·661	3·674	3·688	3·701	3·715	3·728	1	3	4	5	7	8	10	11	12
14	3·742	3·755	3·768	3·782	3·795	3·808	3·821	3·834	3·847	3·860	1	3	4	5	7	8	9	11	12
15	3·873	3·886	3·899	3·912	3·924	3·937	3·950	3·962	3·975	3·987	1	3	4	5	6	8	9	10	11
16	4·000	4·012	4·025	4·037	4·050	4·062	4·074	4·087	4·099	4·411	1	2	4	5	6	7	9	10	11
17	4·123	4·135	4·147	4·159	4·171	4·183	4·195	4·207	4·219	4·231	1	2	4	5	6	7	8	10	11
18	4·243	4·254	4·266	4·278	4·290	4·301	4·313	4·324	4·336	4·347	1	2	3	5	6	7	8	9	10
19	4·359	4·370	4·382	4·393	4·405	4·416	4·427	4·438	4·450	4·461	1	2	3	5	6	7	8	9	10
20	4·472	4·483	4·494	4·506	4·517	4·528	4·539	4·550	4·561	4·572	1	2	3	4	6	7	8	9	10
21	4·583	4·593	4·604	4·615	4·626	4·637	4·648	4·658	4·669	4·680	1	2	3	4	5	6	8	9	10
22	4·690	4·701	4·712	4·722	4·733	4·743	4·754	4·764	4·775	4·785	1	2	3	4	5	6	7	8	9
23	4·796	4·806	4·817	4·827	4·837	4·848	4·858	4·868	4·879	4·889	1	2	3	4	5	6	7	8	9
24	4·899	4·909	4·919	4·930	4·940	4·950	4·960	4·970	4·980	4·990	1	2	3	4	5	6	7	8	9
25	5·000	5·010	5·020	5·030	5·040	5·050	5·060	5·070	5·079	5·089	1	2	3	4	5	6	7	8	9
26	5·099	5·109	5·119	5·128	5·138	5·148	5·158	5·167	5·177	5·187	1	2	3	4	5	6	7	8	9
27	5·196	5·206	5·215	5·225	5·235	5·244	5·254	5·263	5·273	5·282	1	2	3	4	5	6	7	8	9
28	5·292	5·301	5·310	5·320	5·329	5·339	5·348	5·357	5·367	5·376	1	2	3	4	5	6	7	7	8
29	5·385	5·394	5·404	5·413	5·422	5·431	5·441	5·450	5·459	5·468	1	2	3	4	5	5	6	7	8
30	5·477	5·486	5·495	5·505	5·514	5·523	5·532	5·541	5·550	5·559	1	2	3	4	4	5	6	7	8
31	5·568	5·577	5·586	5·595	5·604	5·612	5·621	5·630	5·639	5·648	1	2	3	3	4	5	6	7	8
32	5·657	5·666	5·675	5·683	5·692	5·701	5·710	5·718	5·727	5·736	1	2	3	3	4	5	6	7	8
33	5·745	5·753	5·771	5·779	5·779	5·788	5·797	5·805	5·814	5·822	1	2	3	3	4	5	6	7	8
34	5·831	5·840	5·848	5·857	5·865	5·874	5·882	5·891	5·899	5·908	1	2	3	3	4	5	6	7	8
35	5·916	5·925	5·933	5·941	5·950	5·958	5·967	5·975	5·983	5·992	1	2	2	3	4	5	6	7	8
36	6·000	6·008	6·017	6·025	6·033	6·042	6·050	6·058	6·066	6·075	1	2	2	3	4	5	6	7	7
37	6·083	6·091	6·099	6·107	6·116	6·124	6·132	6·140	6·148	6·156	1	2	2	3	4	5	6	7	7
38	6·164	6·173	6·181	6·189	6·197	6·205	6·213	6·221	6·229	6·237	1	2	2	3	4	5	6	6	7
39	6·245	6·253	6·261	6·269	6·277	6·285	6·293	6·301	6·309	6·317	1	2	2	3	4	5	6	6	7
40	6·625	6·332	6·340	6·348	6·356	6·364	6·372	6·380	6·387	6·395	1	2	3	3	4	5	6	6	7
41	6·403	6·411	6·419	6·427	6·434	6·442	6·450	6·458	6·465	6·473	1	2	2	3	4	5	5	6	7
42	6·481	6·488	6·496	6·504	6·512	6·519	6·527	6·535	6·542	6·550	1	2	2	3	4	5	5	6	7
43	6·557	6·565	6·573	6·580	6:588	6·595	6·603	6·611	6·618	6·626	1	2	2	3	4	5	5	6	7
44	6·633	6·641	6·648	6·656	6·663	6·671	6·678	6·686	6·693	6·701	1	2	2	3	4	5	5	6	7
45	6·708	6·716	6·723	6·731	6·738	6·745	6·753	6·760	6·768	6·775	1	1	2	3	4	4	5	6	7
46	6·782	6·790	6·797	6·804	6·812	6·819	6·826	6·834	6·841	6·848	1	1	2	3	4	4	5	6	7
47	6·856	6·863	6·870	6·877	6·885	6·892	6·899	6·907	6·914	6·921	1	1	2	3	4	4	5	6	7
48	6·928	6·935	6·943	6·950	6·957	6·964	6·971	6·979	6·986	6·993	1	1	2	3	4	4	5	6	6
49	7·000	7·007	7·014	7·021	7·029	7·036	7·043	7·050	7·057	7·064	1	1	2	3	4	4	5	6	6
50	7·071	7·078	7·085	7·092	7·099	7·106	7·113	7·120	7·127	7·134	1	1	2	3	4	4	5	6	6
51	7·141	7·148	7·155	7·162	7·176	7·176	7·183	7·190	7·197	7·204	1	1	2	3	4	4	5	6	6
52	7·211	7·218	7·225	7·232	7·239	7·246	7·253	7·259	7·266	7·273	1	1	2	3	3	4	5	6	6
53	7·280	7·287	7·294	7·301	7·308	7·314	7·321	7·328	7·335	7·342	1	1	2	3	3	4	5	5	6
54	7·348	7·355	7·362	7·369	7·376	7·382	7·396	7·396	7·403	7·409	1	1	2	3	3	4	5	5	6

SQUARE ROOTS

	0	1	2	3	4	5	6	7	8	9	Mean Differences								
											1	2	3	4	5	6	7	8	9
55	7·410	7·423	7·430	7·436	7·443	7·450	7·457	7·463	7·470	7·477	1	1	2	3	3	4	5	5	6
56	7·483	7·490	7·497	7·503	7·510	7·517	7·523	7·530	7·537	7·543	1	1	2	3	3	4	5	5	6
57	7·550	7·556	7·563	7·570	7·576	7·583	7·589	7·596	7·603	7·609	1	1	2	3	3	4	5	5	6
58	7·616	7·622	7·629	7·635	7·642	7·649	7·655	7·662	7·668	7·675	1	1	2	3	3	4	5	5	6
59	7·681	7·688	7·694	7·701	7·707	7·714	7·720	7·727	7·733	7·740	1	1	2	3	3	4	4	5	6
60	7·746	7·752	7·759	7·765	7·772	7·778	7·785	7·791	7·797	7·804	1	1	2	3	3	4	4	5	6
61	7·810	7·817	7·823	7·829	7·836	7·842	7·849	7·855	7·861	7·868	1	1	2	3	3	4	4	5	6
62	7·874	7·880	7·887	7·893	7·899	7·906	7·912	7·918	7·925	7·931	1	1	2	3	3	4	4	5	6
63	7·937	7·944	7·950	7·956	7·962	7·969	7·975	7·981	7·987	7·994	1	1	2	3	3	4	4	5	6
64	8·000	8·006	8·012	8·019	8·025	8·031	8·037	8·044	8·050	8·056	1	1	2	2	3	4	4	5	6
65	8·062	8·068	8·075	8·081	8·087	8·093	8·099	8·106	8·112	8·118	1	1	2	2	3	4	4	5	6
66	8·124	8·130	8·136	8·142	8·149	8·155	8·161	8·167	8·173	8·179	1	1	2	2	3	4	4	5	5
67	8·185	8·191	8·198	8·204	8·210	8·216	8·222	8·228	8·234	8·240	1	1	2	2	3	4	4	5	5
68	8·246	8·252	8·258	8·264	8·270	8·276	8·283	8·289	8·295	8·301	1	1	2	2	3	4	4	5	5
69	8·307	8·313	8·319	8·325	8·331	8·337	8·343	8·349	8·355	8·361	1	1	2	2	3	4	4	5	5
70	8·367	8·373	8·379	8·385	8·390	8·396	8·402	8·408	8·414	8·420	1	1	2	2	3	4	4	5	5
71	8·426	8·432	8·438	8·444	8·450	8·456	8·462	8·468	8·473	8·479	1	1	2	2	3	4	4	5	5
72	8·485	8·491	8·497	8·503	8·509	8·515	8·521	8·526	8·532	8·538	1	1	2	2	3	3	4	5	5
73	8·544	8·550	8·556	8·562	8·576	8·573	8·579	8·585	8·591	8·597	1	1	2	2	3	3	4	5	5
74	8·602	8·608	8·614	8·620	8·626	8·631	8·637	8·643	8·649	8·654	1	1	2	2	3	3	4	5	5
75	8·660	8·666	8·672	8·678	8·683	8·689	8·695	8·701	8·706	8·712	1	1	2	2	3	3	4	5	5
76	8·718	8·724	8·729	8·735	8·741	8·746	8·752	8·758	8·764	8·769	1	1	2	2	3	3	4	5	5
77	8·775	8·781	8·786	8·792	8·798	8·803	8·809	8·815	8·820	8·826	1	1	2	2	3	3	4	4	5
78	8·832	8·837	8·843	8·849	8·854	8·860	8·866	8·871	8·877	8·883	1	1	2	2	3	3	4	4	5
79	8·888	8·894	8·899	8·905	8·911	8·916	8·922	8·927	8·933	8·939	1	1	2	2	3	3	4	4	5
80	8·944	8·950	8·955	8·961	8·967	8·972	8·978	8·983	8·989	8·994	1	1	2	2	3	3	4	4	5
81	9·000	9·006	9·001	9·017	9·022	9·028	9·033	9·039	9·044	9·050	1	1	2	2	3	3	4	4	5
82	9·055	9·061	9·066	9·072	9·077	9·083	9·088	9·094	9·099	9·105	1	1	2	2	3	3	4	4	5
83	9·110	9·116	9·121	9·127	9·132	9·138	9·143	9·149	9·154	9·160	1	1	2	2	3	3	4	4	5
84	9·165	9·171	9·176	9·182	9·187	9·192	9·198	9·203	9·209	9·214	1	1	2	2	3	3	4	4	5
85	9·220	9·225	9·230	9·236	9·241	9·247	9·252	9·257	9·263	9·268	1	1	2	2	3	3	4	4	5
86	9·274	9·279	9·284	9·290	9·295	9·301	9·306	9·311	9·317	9·322	1	1	2	2	3	3	4	4	5
87	9·327	9·333	9·338	9·343	9·349	9·354	9·359	9·365	9·370	9·375	1	1	2	2	3	3	4	4	5
88	9·381	9·386	9·391	9·397	9·402	9·407	9·413	9·418	9·423	9·429	1	1	2	2	3	3	4	4	5
89	9·434	9·439	9·445	9·450	9·455	9·460	9·466	9·471	9·476	9·482	1	1	2	2	3	3	4	4	5
90	9·487	9·492	9·497	9·503	9·508	9·513	9·518	9·524	9·529	9·534	1	1	2	2	3	3	4	4	5
91	9·539	9·545	9·550	9·555	9·560	9·566	9·571	9·576	9·581	9·586	1	1	2	2	3	3	4	4	5
92	9·592	9·597	9·602	9·607	9·612	9·618	9·623	9·628	9·633	9·638	1	1	2	2	3	3	4	4	5
93	9·644	9·649	9·654	9·659	9·664	9·670	9·675	9·680	9·685	9·690	1	1	2	2	3	3	4	4	5
94	9·695	9·701	9·706	9·711	9·716	9·721	9·726	9·731	9·737	9·742	1	1	2	2	3	3	4	4	5
95	9·747	9·752	9·757	9·762	9·767	9·772	9·778	9·783	9·788	9·793	1	1	2	2	3	3	4	4	5
96	9·798	9·803	9·808	9·813	9·818	9·823	9·829	9·834	9·839	9·844	1	1	2	2	3	3	4	4	5
97	9·849	9·854	9·859	9·864	9·869	9·874	9·879	9·884	9·889	9·894	1	1	1	2	3	3	4	4	5
98	9·899	9·905	9·910	9·915	9·920	9·925	9·930	9·935	9·940	9·945	1	1	1	2	2	3	3	4	4
99	9·950	9·995	9·960	9·965	9·970	9·975	9·980	9·985	9·990	9·995	0	1	1	2	2	3	3	4	4

Reproduced here from Frank Castle's *Four-Figure Mathematical Tables,* Macmillan

ANSWERS

Exercises 1.1 (page 12)

1) 16 200
2) 46·1
3) 4 216 310
4) 10·2
5) 760
6) 0·525
7) 32·761
8) 52
9) 0·043 72
10) 0·005 04
11) 43 100
12) 52·1
13) 620
14) 23
15) 6
16) 0·0134
17) 0·2326
18) 0·017 415
19) 70
20) 0·052

Exercise 1.2 (page 18)

1) 0·7
2) 0·093 75
3) $0{\cdot}1\dot{6}$
4) $0{\cdot}0\dot{6}$
5) 13·57
6) 83·048
7) 652·1622
8) 37·535
9) 43·5
10) 14·29
11) 4·66
12) 13·656
13) 38·01
14) 8·0115
15) 0·347 976
16) 6679·742
17) 1720
18) 374
19) 6·31
20) 0·017

Exercise 1.3 (page 21)

1) a) kilogramme b) second c) metre
2) a) 0·35 b) 6·24 c) 240 d) 6000 e) 0·027 f) 625 g) 25 h) 30 i) 700 j) 102
3) a) 3 b) 5·2 c) 3000 d) 22 e) 3·562 f) 220 g) 560 h) 220 i) 0·51 j) 0·02
4) 17·435
5) £3·88
6) 562·74 mm
7) 0·076 mm or 76 μm
8) 39
9) £58·975
10) 2·61 m
11) 11·487 mm
12) 1200
13) 400 kg
14) £3·46

Exercise 1.4 (page 28)

1) 506
2) 21·86
3) 5·33
4) 9·1; 11
5) 60
6) 936 km/litre
7) a) 14 b) 13
8) 1·4
9) 4:7

10) 42:57
11) 5:4
12) 1:40
13) 10:3
14) 1:400
15) 160 mm
16) 2·34 Mm
17) 54 Mm
18) 625
19) £105
20) £24; £30; £36

Exercises 1.5 (page 31)

1) 0·025
2) 0·055
3) 0·26
4) 0·025
5) 60%
6) 18%
7) 128%
8) 21·7%
9) 0·1875 l
10) 0·14 l
11) 0·0495 m
12) 1·3125 kg
13) 1·69 l
14) 9·6 l
15) a) 25%
b) 1:3
16) 40 Mm
17) £12·42
18) £9·375
19) 4
20) 8%

Exercise 1.6 (page 33)

1) 43·2
2) 40
3) 0·56
4) 1·524
5) 226·8
6) 623·7
7) 34
8) 4·5
9) 48
10) 5·3
11) 1 mm
12) 89·15
13) 49·95
14) 36·29 kg @ 7·94 mm
15) 32 km
16) 12 mm
17) 70·5 km
18) 3·048 tonne
19) 2000
20) 28·4 hours

Exercise 1.7 (page 41)

1) a) 0·455 m^2
b) 3·3 m
2) 3·3 km
3) 5000
4) 5·544 cm^3
5) 27·3 cm^2
6) 75 l
7) 3·375
8) 422·4 cm^3
9) a) 138 mm
b) 1386 mm^2
10) a) 0·54 m^2
b) 27 l

Exercise 1.8 (page 50)

1) 35°C
2) 9·5 ohms
3) 955
4) 350 cm^3
5) 32 m
6) 13b–c
7) $6y - 2x$
8) $2a - b$
9) $14y - 2x$
10) $28a^5$
11) $-21x^6y^3$
12) $4a^3$
13) $6xy^2$
14) 3
15) 5
16) 3
17) $\frac{L - l}{lt}$
18) $\frac{4v}{\pi d^2}$
19) $\sqrt{\frac{4v}{\pi h}}$
20) $\frac{E}{R}$
21) $\sqrt{c^2 - a^2}$

Exercise 1.9 (page 53)

1) 1080°
2) 5
3) 240°
4) 33 mm
5) induction
6) 30°; 120°; 225°

Exercise 1.10 (page 56)

1) a) 5:1 b) 500 rev/min
2) 9
3) 12:1 & 300 rev/min
4) a) 2·1:1 b) 2000 rev/min
5) 24
6) 2·1:1
7) 1806 rev/min
8) 90°

Exercise 1.11 (page 61)

1) 4·359
2) 5·215
3) 2·499
4) 9·933
5) 27·04
6) 52·41
7) 66·01
8) 0·2704
9) 7·81
10) 3·852
11) 6·675
12) 8·906
13) 96·83 mm
14) 28·17 mm
15) 206·2 mm
16) 2·102 m
17) 8 mm

Exercise 1.12 (page 64)

1) 9·2 newtons
2) 9 V
3) 18 kW
4) 4·472
5) 110°C

Exercise 1.13 (page 66)

1) 1273
2) 1591
3) 4773
4) 1909
5) 1045
6) 0·24 mm/rev & 636
7) 0·5
8) yes

Exercise 1.14 (page 73)

1) 7·12 mm
2) 12·59 mm
3) 18·87 mm
4) 0·5 mm

Exercise 2.7 (page 90)

1c	2c	3b	4b	5b	6b	7d	8c	9a	10d
11c	12b	13b	14d	15c	16b	17c	18a	19a	20c
21c	22a	23c	24a	25c	26d	27c	28d	29a	30d
31b	32c	33c	34a	35d	36d	37c	38d	39b	40c

Exercise 2.13 (page 104)

1a	2a	3b	4b	5c	6c	7b	8d	9c	10d
11c	12d	13d	14c	15b					

Exercise 2.17 (page 111)

1a	2b	3a	4a	5d	6a	7d	8c	9c	10a	
11a	12b	13c	14c	15c	16c	17a	18b	19c	20d	21b

Exercise 2.19 (page 121)

1d	2a	3b	4d	5c	6a	7d	8c	9c	10d
11a	12b	13b	14d	15a	16d	17d	18c	19c	20c
21c	22d	23d	24b						

Exercise 2.25 (page 133)

1a	2b	3c	4d	5b	6a	7b	8d	9b	10b
11a	12a	13a	14a	15b	16b	17b	18d	19d	20a
21b	22a	23c	24a	25c					

Exercise 2.34 (page 145)

1a	2b	3c	4d	5a	6b	7d	8a	9d	10a
11c	12c	13d	14b	15b	16d	17d	18a	19b	20d
21c	22b	23d	24b	25d	26a	27a	28d	29c	30b
31d	32a	33d	34b	35b	36c				

Exercise 2.38 (page 154)

1c	2d	3b	4a	5d	6b	7c	8b	9c	10d
11c	12a	13b	14d	15c	16b	17c	18b	19a	20a

Exercise 3.10 (page 171)

1d	2d	3a	4d	5a	6c	7b	8d	9a	10a
11a	12d	13b	14b	15d	16d	17c	18b	19d	20a
21c	22b	23b	24c	25a	26b				

INDEX

Conversion – Imperial to SI 31, 176